The book presents, in a condensed form, all aspects of special relativity (part I) and relativistic gravitation (part II). The emphasis is on experiment, observation and physical interpretation.

The first four chapters in part I introduce special relativity, the following three chapters deal with mathematical background and introduce new concepts useful for part II.

The first five chapters in part II describe Newtonian gravitation as well as metric theories (including general relativity); experimental tests are described using the post-Newtonian parameters; the Newtonian and relativistic cosmologies are presented. The following three chapters contain detailed discussion of general relativity by means of the usual geometric tools introduced for this purpose. The last chapter presents two practical projects: the satellite gyroscope to measure the gravitational interactions spin–orbit and spin–spin, and a gravitational wave detector.

Relativity and Gravitation

Relativity and Gravitation

Philippe Tourrenc

Université Pierre et Marie Curie

Translated by

Andrew King

University of Leicester

CAMBRIDGE
UNIVERSITY PRESS

PUBLISHED BY THE PRESS SYNDICATE OF THE UNIVERSITY OF CAMBRIDGE
The Pitt Building, Trumpington Street, Cambridge CB2 1RP, United Kingdom

CAMBRIDGE UNIVERSITY PRESS
The Edinburgh Building, Cambridge CB2 2RU, United Kingdom
40 West 20th Street, New York, NY 10011-4211, USA
10 Stamford Road, Oakleigh, Melbourne 3166, Australia

First published in French as *Relativité et gravitation*
by Armand Colin Editeur, Paris, 1992, and © Armand Colin Editeur, 1992

First published in English by Cambridge University Press, 1997
English translation © Cambridge University Press, 1997

Printed in the United Kingdom at the University Press, Cambridge

Typeset in Times 10/12pt

A catalogue record for this book is available from the British Library

Library of Congress Cataloguing in Publication Data

Tourrenc. Philippe.
[Relativité et gravitation. English]
Relativity and gravitation / Philippe Tourrenc ; translated by
Andrew King
p. cm.
Includes bibliographical references and index.
ISBN 0 521 45075 6. – ISBN 0 521 45685 1 (pbk.)
1. Special relativity (Physics) 2. Relativity (Physics)
3. Cosmology. I. Title.
QC173.65. T6813 1997
530.1′1–dc20   96-2859 CIP

ISBN 0 521 45075 6 hardback
ISBN 0 521 45685 1 paperback

Contents

Preface

This book is *an introduction to relativistic gravitation: the special and general theories of relativity*, which we can regard as members of the larger class of *metric theories of gravity*. The same formalism is equally adapted for treating *inertial effects*. In this treatment we shall highlight the choices made in constructing the theory, and where appropriate point out their originality as compared with the many current treatments, particularly of special relativity.

The foundations of special relativity were laid in 1905 in a famous paper by Albert Einstein "On the Electrodynamics of Moving Bodies" ("Zur Elektrodynamik bewegter Körper", *Annalen der Physik,* **17**). General relativity was complete by the end of 1915. In 1916 Einstein published "The Foundations of the General Theory of Relativity" ("Die Grundlage der allgemeinen Relativitätstheorie", *Annalen der Physik,* **49**) in which he presented the theory and its first results.

The achievement was enormous: impressive in its clarity of argument and its novelty of ideas, in its power of prediction, in its originality and its depth and concision of thought.

But these papers were also in some ways misleading, as a decisive break was needed with the Newtonian picture of an absolute space with a frozen geometry, in which objects "aged" following the flow of a universal time.

The Newtonian concepts of space, time and force are profoundly rooted in our consciousness through the daily use of language making them familiar to all of us. However, relativity theory denies some of them any intrinsic physical reality. This makes relativity theories inherently difficult.

Einstein did not overcome the difficulties caused by poorly adapted language in either of the two papers cited above. In the first of them Einstein still used the habitual concepts of time and length, although attributing new properties to them. In the second paper he posed the problem of the behaviour of clocks with some confusion of language and possibly of thought. According to this paper, "clocks run more slowly in the vicinity of a gravitating mass." This interpretation implicitly assumes the existence of some unknown reference chronology somehow allowing us to discover changes in the behaviour of clocks. This interpretation is quite contrary to current understanding of general relativity and in conflict with the positivism inspiring the analysis of simultaneity in 1905.

Of course, such difficulties could not be overcome instantly; time was needed to absorb and penetrate these profoundly novel theories. However, even today some textbooks and research papers manifest these same confusions, and spread immense misunderstanding. This arises because special relativity is an indispensable theory in many fields, while general relativity is still the preserve of a comparatively small group of physicists. An independent neo-Newtonian interpretation has therefore arisen as a direct continuation of Einstein's 1905 paper. This interpretation is of no use for general relativity, even though the conceptual basis of the two theories is the same.

From all the ways of introducing special relativity, we have chosen the one leading naturally to general relativity. We do not claim at all that it is necessary systematically to "speak relativistically", and to be precise but incomprehensible; usually all that is necessary is to know what relativistic reality is being described in Newtonian language in a given case.

Observational tests of general relativity have become common and in 1996 there is no room to doubt the theory. Moreover astrophysical observational techniques and experimental possibilities are evolving rapidly. One can easily envisage new experiments or the repetition of older ones with greater sensitivity. General relativity is no longer purely a theory, resting on "the three classical tests" of thirty years ago. Relativistic properties of gravitation and inertia interest a growing set of physicists, although the corresponding theories do not yet form part of the standard training of a physicist. It would be regrettable if this lack retarded a trend which is already under way in some quarters. *This book is therefore addressed not only to beginners or near-beginners, but also to physicists specialising in other areas.*

The starting level of this book is that of an undergraduate university physics course, even if not all the subjects covered form part of the course.

The book does not aim to be exhaustive as there are many very good complete books in the literature (see the bibliography at the end of the book). Our intention is to explain the physical nature of relativity theories without hiding the difficulties which are frequently passed over in silence, and to allow access to the specialist literature.

The book is intended as a *physics text*, rather than one full of mathematical derivations which would confine the readership to specialists, or require too much preparatory work. In particular we have avoided using the general, powerful and elegant methods of differential geometry. Further, *we have implicitly assumed that all the conditions required for the various proofs are satisfied*, for example that all the functions introduced have the regularity properties needed. Nevertheless we have rigorously adhered to the concepts we regard as fundamental, even at the price of sometimes being fastidious.

When a physicist explores the properties of space-time experimentally, he determines the positions and times of events by measurements followed by calculations. He thus finds the "coordinates of the events". He rarely asks what another physicist moving with respect to him would find. On the contrary, his concern is to compare his observations with theoretical predictions, either to

CERN, the European Centre for Nuclear Research, is a vast complex of several accelerators at Geneva. It is mainly a European laboratory for studying fundamental particles.

Situated astride the Franco-Swiss frontier, the LEP (Large Electron Positron collider) has a diameter of about 9 km; it can reach particle energies of the order of 50 GeV.

The diameter of the SPS (Super Proton Synchrotron) is of the order of 2 km. The SPS accelerates protons and antiprotons from 28 GeV to 450 GeV in about ten seconds.

The two rings marked on the photograph above can be used as colliders (cf. § 4.4.4).

As an exercise the reader should calculate how the radius of the rings varies as a function of the mass and energy of the particles (cf. § 4.4.3).

(Photograph: CERN)

test the theory, or to use it to find the numerical value of some quantity. For such purposes it is often unnecessary to know whether there exist "coordinate systems such that...". It is much more important to know the theoretical predictions in the coordinate system actually being used. For this reason we have chosen to give the Lorentz transformation no more prominence than that traditionally given to the Galilean transformation in an introduction to Newtonian mechanics. Similarly in presenting theories of gravitation we shall emphasise the role of coordinates rather than a geometrical approach to the intrinsic properties of space-time. These choices involve a certain shift between the physical ideas expressed and the vocabulary used. Looking a little deeper we see that the viewpoints of the "physicist" and "geometer" are complementary. Our standpoint nevertheless requires us to *pay particular attention to the physical interpretation of the theory.*

In this spirit we have tried to produce a book in which *only the bases of the Newtonian theory of dynamics and gravitation are assumed at the outset.*

Readers of scientific books do not always begin at the beginning. To allow easy access to the various subjects we give an overview at the beginning of each chapter and give as many reminders as possible of notation, properties derived earlier, and so on.

To help the understanding of essentials, and to avoid adding technical difficulties to conceptual ones, we have put into smaller type material that can or should be omitted on a first reading, such as long proofs for example.

To understand a theory, one has to distinguish clearly between the mathematical framework and the physical properties it is designed to describe. To avoid possible ambiguity, the symbol "≡" is sometimes used to denote definitions or properties deduced from them, in addition to its usual meaning. This is in contrast to the symbol "=" which is used to express physical laws and equations expressing them.

The sections and equations are numbered. Unless explicitly mentioned, references are to sections or equations of the same part of the book.

Among those who are not quoted directly in this book, it is a pleasure to thank Mmes Man and Ouannes and M. Le Denmat as well as my other colleagues, physicists and secretaries, of the Laboratoire de Physique de l'Institut Henri-Poincaré (UPMC) for the help they have given me, sometimes without knowing.

I want to thank, especially, Prof. Andrew King from Leicester University for the translation of this book.

I

Special Relativity

Chapter 1

Kinematics of a point mass

Every physical theory assumes a descriptive framework and a set of interpretations. To a large extent, Newtonian and relativistic mechanics involve identical concepts. The common descriptive framework of the two types of mechanics is founded on the notion of an event.

An **event** is defined by its nature: a light is switched on, a nucleus disintegrates, two trains pass each other, etc.

It is specified by four coordinates; suitably chosen, these can be interpreted precisely in both theories.

In this chapter we shall present the bases of Newtonian mechanics and Special Relativity in such a way as to allow direct comparison. We assume the reader is familiar with Newtonian theory, and pay special attention to the geometrical framework of relativity.

1.1 The basis of Newtonian kinematics

Newtonian mechanics assumes the existence of an **absolute space** endowed with physical reality, in which events occur. In mathematical terms this space is Euclidean and affine. *The distance between two points is a physical quantity whose existence is postulated a priori, independently of any way of measuring it.* The accurate measurement and comparison of distances is a problem for experimental physics.

Points in this space are specified by three coordinates, x, y, z. These may be chosen so that the distance, ℓ, between two points A and B can be expressed as

$$\ell = \sqrt{(x_A - x_B)^2 + (y_A - y_B)^2 + (z_A - z_B)^2},$$

(1.1.1)

where x, y, z and ℓ are measured in metres.

In reality, only those measurements actually performed are accessible to us. Asserting the Euclidean character of space is a way of summarising the relations between measurements of real objects. This assertion results in a highly-developed theory, which was so well confirmed experimentally that there could be no doubt of its validity for centuries.

Given the spatial coordinates of the point where the event occurred, a fourth coordinate, t, specifies the time of the event. This time coordinate can be chosen so as to have a physical meaning.

We consider a system of small size, moved from A at the time t_A (event E_A) to B at the time t_B (event E_B).

This physical system "ages", i.e. suffers an irreversible evolution: a set of cells reproduces, radioactive nuclei disintegrate, the small hand of a clock turns a number of times noted by an observer, etc. This irreversibility allows us to state unambiguously that for example the system was at A **before** B.

The idea of time-ordering is independent of the physical system considered and the way in which the system is moved from A to B: it depends only on the events A and B considered.

Time-ordering is a transitive relation in the set of events: E_A before E_B and E_B before E_C implies E_A before E_C, whatever the events E_A, E_B and E_C considered.

The "ageing" of the system is measured by a positive quantity: **its proper time** τ_{AB}. *In the specific framework of Newtonian theory, the time* τ_{AB} *depends only on the two events* E_A *and* E_B *considered.* An absolute duration is thus associated *a priori* with the two events E_A and E_B. The time coordinate t may therefore be chosen so that

$$\tau_{AB} = t_B - t_A ; \tag{1.1.2}$$

here t is in seconds.

The time t is called **absolute time**; the set of events characterised by the same value of t constitutes an **instant** E_t, which we can identify with absolute space.

The motion of a point mass M is characterised by the fact that its spatial coordinates vary with time; it is thus specified by three functions X, Y, Z:

$$x = X(t), \quad y = Y(t), \quad z = Z(t).$$

It is convenient to use a space–time diagram (fig. 1.1) to represent this motion. The motion is then described by a curve. If the moving body is extended, the set of curves describing the motion of each of its points constitutes a tube.

Let us consider two moving points, M and P. At a given instant they are at the points $X_M(t), Y_M(t), Z_M(t)$ and $X_P(t), Y_P(t), Z_P(t)$ of coordinate space. Their distance at this instant is

$$\ell = \sqrt{\{X_M(t) - X_P(t)\}^2 + \{Y_M(t) - Y_P(t)\}^2 + \{Z_M(t) - Z_P(t)\}^2}. \tag{1.1.3}$$

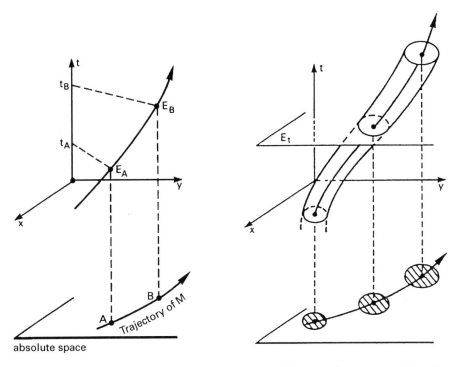

Motion of a point mass   Motion of an extended body

Fig. 1.1

The absolute system of coordinates we have introduced is not the only one possible. We can specify the events by constructing new coordinates, setting

$$x' = F(x, y, z, t), \quad y' = G(x, y, z, t), \quad z' = H(x, y, z, t), \quad t' = S(x, y, z, t),$$

where the functions F, G, H and S are such that these relations are invertible, and allow us to express x, y, z, t as functions of x', y', z' and t'. This is required so as to avoid any ambiguity in the specification of events.

We shall designate as **Newtonian coordinate systems** those systems satisfying (1.1.2) and (1.1.3).

The existence of Newtonian coordinate systems provides the basis for Newtonian kinematics. The absolute system is a Newtonian coordinate system, but not the only one (cf. the next chapter).

1.2 The basis of relativistic kinematics

In relativistic theories, the idea of an event also plays a central role. Here too, events are *specifiable* by four coordinates x, y, z, t. However, we can no longer distinguish space from time in an intrinsic fashion. Only the full set of events has physical meaning: this set constitutes the **space-time**. The concept of "where" an event occurred has no physical meaning; similarly the concept of "when" the event occurred. The notion of distance as an *a priori* defined quantity disappears, along with the concept of an instant.

All that remains is the concept of ageing, along with the notions of time-ordering and proper time which flow from it.

The motion of a pointlike body is described by a curve in space-time. This curve is not a simple diagrammatic representation, but has an intrinsic physical reality; this curve is the **worldline** or space-time trajectory of the body. The motion of an extended body is represented by a **world-tube**.

To represent events and motion of bodies we need a coordinate system. At this point the representation becomes diagrammatic and is identical to the corresponding Newtonian description except for the physical interpretation of the reality represented. The type of representation stays unchanged, so fig. 1.1 also describes the situation considered. However the interpretation is profoundly changed.

Let us consider a space–time diagram and two events E_1 and E_2, on the worldline of a body. In the Newtonian case the ageing of the body between E_1 and E_2 is independent of the space–time trajectory between the two events. For many years, experimental and observational accuracy was insufficient to raise doubts. *We now know that the ageing depends on the way in which the body moves between E_1 and E_2.*

Let us assume that E_1 and E_2 are neighbouring events on the worldline of a certain body, with coordinates (x, y, z, t) and $(x + dx, y + dy, z + dz, t + dt)$ respectively.

Special relativity postulates that there exists a system of privileged coordinates such that the proper time $d\tau$ between these two events has the form

$$d\tau = \sqrt{\left\{ dt^2 - \frac{dx^2 + dy^2 + dz^2}{c^2} \right\}}. \tag{1.2.1}$$

Such coordinates define a **Minkowski coordinate system**; it is not unique. In the expression for the proper time $d\tau$, c is a universal constant with the dimensions of a velocity (m s^{-1}), whose interpretation we shall give later.

Consider the worldline of a body between two events which are not necessarily close, with coordinates (x_1, y_1, z_1, t_1) and (x_2, y_2, z_2, t_2) respectively. This worldline is defined by the three functions

$$x = X(t), \quad y = Y(t), \quad z = Z(t),$$
$$x_k = X(t_k), \quad y_k = Y(t_k), \quad z_k = Z(t_k), \quad \text{with } k = 1 \text{ or } 2.$$

The ageing of the body between these two events is measured by its proper time τ_{12}, defined as the sum of proper time elements

$$\tau_{12} = \int_{t_1}^{t_2} \sqrt{1 - \beta^2}\, dt \quad \text{for } t_2 > t_1 \tag{1.2.2}$$

where we have set $\beta^2 = \{(dX/dt)^2 + (dY/dt)^2 + (dZ/dt)^2\}/c^2$.

Several remarks are in order concerning this formula:

(1) When β^2 is small compared with unity the expression for τ_{12} becomes the Newtonian expression: $\tau_{12} = t_2 - t_1$. This suggests that the Minkowski coordinates used become Newtonian coordinates if the bodies under consideration move so that $\beta^2 << 1$; this situation is the "low-velocity limit".

(2) Quite generally, one can define a *Newtonian language* once the Minkowski coordinate system is fixed.

We call the set of events with coordinate t an "instant". We can identify the set E_t of events at a given instant with the Euclidean space in Newtonian theory. We identify t with absolute time. This allows us to *define* the distance between two points of the space E_t by use of the expression (1.1.1) and to introduce the notion of velocity. However, we must remember that these definitions are relative to a Minkowski coordinate system. There are many Minkowski coordinate systems (cf. the next chapter), and we will see that the Newtonian language is simply a convenience, which does not express any intrinsic physical reality.

(3) In this language, β appears as the ratio V/c, where V is the "velocity" of the body.

We may summarise the situation as follows:

"Relativity gives the same results as Newtonian theory in the low-velocity limit."

(4) The most remarkable consequence of the relativistic expression for proper time is the existence of a limiting velocity c. We see that for $d\tau$ to exist and allow us to calculate τ_{12}, the relation $\beta^2 \leqslant 1$ must hold. This shows that not all events can be linked by the motion of a body between them.

In fig. 1.2, we consider two particular events A and B in the (x, t) plane, and a worldline in this plane which joins the two events: this has $\beta^2 > 1$ near N. A worldline like this can never represent the motion of a physical particle.

1.3 Minkowski space-time

In special relativity, the ensemble of all events constitutes a four-dimensional pseudo-Euclidean *affine space*: Minkowski space-time. The ensemble of points of Newtonian space constitutes a three-dimensional Euclidean affine space, which explains the geometrical analogies we shall encounter.

1.3.1 Mathematical basis of Minkowski geometry

The geometrical bases of special relativity introduce two distinct spaces:

(a) A space of points M_4, describable by four coordinates: this is the set of events.

(b) A four-dimensional vector space, T_4, constructed from the real numbers.

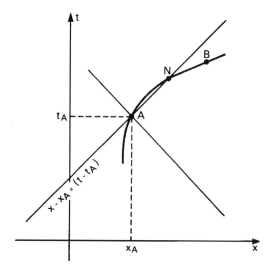

Fig. 1.2

We recall briefly the structure of the **affine space** M_4:

(1) To each pair of points (A, B) of M_4 corresponds a vector of T_4, called $\overrightarrow{AB}$, such that

$$\overrightarrow{AB} = -\overrightarrow{BA} \quad \text{and} \quad \overrightarrow{AB} + \overrightarrow{BC} + \overrightarrow{CA} = \overrightarrow{0}$$

for any A, B and C belonging to M_4.

(2) Choosing a point O of M_4 once and for all, each vector $\overrightarrow{U}$ of T_4 corresponds to exactly one point A such that $\overrightarrow{OA} = \overrightarrow{U}$.

We define a **frame** by choosing a point O of M_4 and a basis $\{\vec{e}_\alpha\}$ of T_4 ($\alpha = 0, 1, 2, 3$). An arbitrary point A of M_4 is then determined by the components x^α of the vector $\overrightarrow{OA}$:

$$\overrightarrow{OA} = \sum_\alpha x^\alpha \cdot \vec{e}_\alpha.$$

The four numbers $\{x^\alpha\} = \{x^0, x^1, x^2, x^3\}$ are the **natural coordinates** of A in the frame $\{0, \vec{e}_\alpha\}$; conversely, the basis $\{\vec{e}_\alpha\}$ is called the **natural basis** associated with the coordinates $\{x^\alpha\}$.

The pseudo-Euclidean structure is introduced by defining a bilinear form in T_4, called the **scalar product**.

With every pair of vectors $(\overrightarrow{U}, \overrightarrow{V})$ of T_4, we associate a real number $(\overrightarrow{U}, \overrightarrow{V})$ such that

$$(\overrightarrow{U}, \overrightarrow{V}) = (\overrightarrow{V}, \overrightarrow{U}) \quad \text{and} \quad (\overrightarrow{U}, \overrightarrow{V} + \mu \cdot \overrightarrow{W}) = (\overrightarrow{U}, \overrightarrow{V}) + \mu \cdot (\overrightarrow{U}, \overrightarrow{W})$$

for any $\overrightarrow{U}, \overrightarrow{V}, \overrightarrow{W}$ belonging to T_4, with μ real.

Further *we postulate the existence of a Minkowski basis, $\{\vec{e}_\alpha\}$, for T_4* such that $(\vec{e}_\alpha, \vec{e}_\beta) = \eta_{\alpha\beta}$, where $\eta_{\alpha\beta}$ is defined by

$$\eta_{\alpha\beta} = 0 \quad \text{for} \quad \alpha \neq \beta$$
$$\eta_{00} = 1 \quad \eta_{kk} = -1 \quad (k = 1, 2, 3).$$

This completely defines the scalar product of any two vectors whose decomposition in the basis $\{\vec{e}_\alpha\}$ is known:

$$\vec{U} = \sum_\alpha X^\alpha \cdot \vec{e}_\alpha \quad \text{and} \quad \vec{W} = \sum_\beta Y^\beta \cdot \vec{e}_\beta$$

giving the expression

$$(\vec{U}, \vec{W}) = X^0 \cdot Y^0 - X^1 \cdot Y^1 - X^2 \cdot Y^2 - X^3 \cdot Y^3.$$

The square of the pseudo-distance between two events A and B is then defined as the scalar product $(\overrightarrow{AB}, \overrightarrow{AB})$; this is the pseudo-norm of $\overrightarrow{AB}$:

$$(\overrightarrow{AB}, \overrightarrow{AB}) = (X_B{}^0 - X_A{}^0)^2 - (X_B{}^1 - X_A{}^1)^2 - (X_B{}^2 - X_A{}^2)^2 - (X_B{}^3 - X_A{}^3)^2.$$

We recover the physical interpretation of the last paragraph by setting

$$X^0 = ct, \quad X^1 = x, \quad X^2 = y, \quad X^3 = z$$

$\sqrt{(\overrightarrow{AB}, \overrightarrow{AB})} = c\tau_{AB}$ for $(\overrightarrow{AB}, \overrightarrow{AB}) > 0$ and A and B close (cf. 1.2.1).

Minkowski bases are not the only ones possible for T_4. Further, once a basis is chosen, nothing forces us to use the associated natural coordinates to specify events. However, the use of an arbitrary coordinate system complicates the notation.

1.3.2 Comparison of Euclidean and Minkowski geometries

Consider the subspace E_t of M_4 defined above; E_t is an affine space. The coordinates are $\{x^k\}$, $k = 1, 2, 3$, and the natural basis is $\{\vec{e}_k\}$. We define a scalar product: $\vec{e}_k \cdot \vec{e}_j = \delta_{kj}$, where δ_{kj} is the Kronecker delta; $\delta_{kj} = 0$ for $k \neq j$ and $\delta_{kk} = 1$ for all k. The introduction of this scalar product allows one to determine the distance ℓ of two events A and B of E_t according to (1.1.1): $\ell^2 = \overrightarrow{AB} \cdot \overrightarrow{AB}$. This mathematical framework gives E_t properties identical to those of absolute Newtonian space.

We note that the mathematical framework of relativity is similar to that which we have just recalled.

Apart from the number of dimensions, the only difference is the form of the scalar product. This is positive definite in three-dimensional Newtonian space. This is not true in relativity.

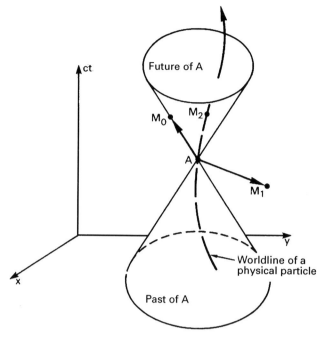

Fig. 1.3

Let us consider two points A and B of E_t. Let $(\overrightarrow{AB}, \overrightarrow{AB})$ be the scalar product obtained by regarding A and B as two events of M_4 and $\overrightarrow{AB} \cdot \overrightarrow{AB}$ the scalar product obtained by regarding A and B as two points of Newtonian space. We get $(\overrightarrow{AB}, \overrightarrow{AB}) = -\overrightarrow{AB} \cdot \overrightarrow{AB}$. To avoid the minus sign, "geometers" prefer to define Minkowski bases by the relation $(\vec{e}_\alpha, \vec{e}_\beta) = f_{\alpha\beta} = -\eta_{\alpha\beta}$. However, some minus signs that the "physicists" would prefer to avoid appear elsewhere. This choice is purely a matter of convention.

1.3.3 The light cone

Consider the event A, given once and for all. Let M be an arbitrary event. The set of events M_0 such that $(\overrightarrow{AM_0}, \overrightarrow{AM_0}) = 0$, constitutes the **light cone** of the point A (fig. 1.3). We shall see later how light is involved in this definition, once the constant c has been interpreted physically (cf. section 2.4).

The light cone divides space-time into various regions.

(1) The set of events M_1 such that $(\overrightarrow{AM_1}, \overrightarrow{AM_1}) < 0$ constitutes the exterior of the cone. No body can move from A to M_1, as its worldline would have to satisfy the relation $\beta^2 > 1$ at at least one of its points, which is not allowed. *There is thus no way of defining a time-ordering between the event M_1 and the event A.*

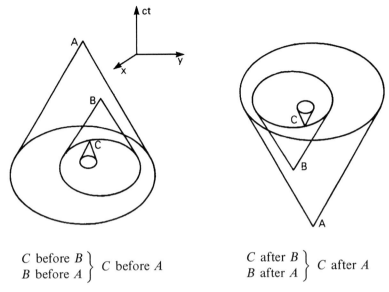

$$C \text{ before } B \left.\vphantom{\begin{matrix}a\\b\end{matrix}}\right\} \; C \text{ before } A \qquad C \text{ after } B \left.\vphantom{\begin{matrix}a\\b\end{matrix}}\right\} \; C \text{ after } A$$
$$B \text{ before } A \qquad\qquad\qquad B \text{ after } A$$

Fig. 1.4

(2) The set of events M_2 such that $(\overrightarrow{AM_2}, \overrightarrow{AM_2}) > 0$ constitutes the interior of the cone. Events A and M_2 can be situated on the worldline of a physical particle. It is thus possible to define time-orderings between A and M_2. The topology of the light-cone is compatible with the transitivity of the time-ordering relations (fig. 1.4), allowing one to distinguish two regions: the **future and past light cones** of A.

It is convenient to introduce the following definitions in T_4:

– a vector $\overrightarrow{U}$ is called **null** iff $\quad (\overrightarrow{U}, \overrightarrow{U}) = 0$,
– it is called **spacelike** iff $\quad (\overrightarrow{U}, \overrightarrow{U}) < 0$,
– and **timelike** iff $\quad (\overrightarrow{U}, \overrightarrow{U}) > 0$.

Of course, had we adopted the convention of the "geometers" for the definition of the scalar product, all the signs would have been reversed.

Given a future-pointing $\vec{e}_0$, a timelike vector is called

– **future-pointing** iff $\quad U^0 = (\overrightarrow{U}, \vec{e}_0) > 0$
– **past-pointing** iff $\quad U^0 = (\overrightarrow{U}, \vec{e}_0) < 0$.

[Note that "iff" means "if and only if".]

1.4 Four-vectors

Quite generally, we define a **four-vector** as a vector of T_4.

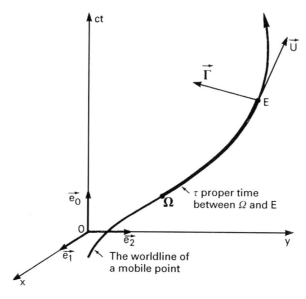

Fig. 1.5

1.4.1 Four-velocity

Consider the representation *in Newtonian theory* of motions in the absolute Newtonian coordinate system $\{x, y, z, t\}$. The trajectory is parametrised by the time t. If the origin is O and the body has position M at time t we define the velocity vector as $\vec{V} = d\overrightarrow{OM}/dt$. The components of $\vec{V}$ in the associated natural basis are $V^1 = dx/dt$, $V^2 = dy/dt$, $V^3 = dz/dt$. Introducing the notation $x^1 = x$, $x^2 = y$, $x^3 = z$, we can write concisely $V^k = dx^k/dt$ with $k = 1, 2, 3$.

Similarly, *we can define the four-velocity of a body in special relativity*.

Consider the worldline of the body in a Minkowski coordinate system. This is a curve we can parametrise using the proper time τ, measured along the curve starting from an arbitrary origin Ω (fig. 1.5). The events E making up the body's worldline have coordinates

$$x \equiv x^1 = X^1(\tau); \quad y \equiv x^2 = X^2(\tau); \quad z \equiv x^3 = X^3(\tau); \quad ct \equiv x^0 = X^0(\tau)$$

with $c^2 d\tau^2 = (dx^0)^2 - (dx^1)^2 - (dx^2)^2 - (dx^3)^2 > 0$.

If the origin is O, we define the four-velocity $\vec{U} = d\overrightarrow{OE}/d\tau$.

Just as in Newtonian theory the velocity $\vec{V}$ is tangent to the body's trajectory, in special relativity the four-velocity is tangent to the body's space-time trajectory.

The components of $\vec{U}$ in the associated natural basis are

$$U^\alpha = \frac{dx^\alpha}{d\tau} \quad \text{with } \alpha = 0, 1, 2, 3. \tag{1.4.1}$$

These are the components of a vector of T_4.

We note that the choice of parameter τ, and the expression $c^2 d\tau^2$ above which follows from it, give the relation

$$\bullet \qquad (\vec{U}, \vec{U}) = (U^0)^2 - (U^1)^2 - (U^2)^2 - (U^3)^2 = c^2. \qquad (1.4.2)$$

In the limit of small velocities $d\tau = dt$, and we can interpret U^k as the component V^k of the Newtonian velocity.

1.4.2 Four-acceleration

At each point on the worldline of a body we can define its four-velocity $\vec{U}$. The vector $\vec{U}$ depends on the point considered, and thus on the parameter τ. We define the four-acceleration, $\vec{\Gamma}$:

$$\bullet \qquad \vec{\Gamma} = \frac{d\vec{U}}{d\tau} = \sum_\alpha \Gamma^\alpha \, \vec{e}_\alpha; \quad \Gamma^\alpha = \frac{dU^\alpha}{d\tau} = \frac{d^2 x^\alpha}{d\tau^2}. \qquad (1.4.3)$$

Differentiating the quantity $(\vec{U}, \vec{U})$ with respect to τ, we find $U^0 \cdot dU^0/d\tau - U^1 \cdot dU^1/d\tau - U^2 \cdot dU^2/d\tau - U^3 \cdot dU^3/d\tau = 0$, or

$$\bullet \qquad (\vec{U}, \vec{\Gamma}) = 0. \qquad (1.4.4)$$

In the limit of small velocities, $dt = d\tau$, we can interpret the components Γ^k as the components of the familiar Newtonian acceleration.

Only three components of the four-velocity and four-acceleration are independent. The situation is analogous to a curve in three-dimensional Euclidean space parametrised by its length ℓ: the tangent vector $\vec{T}$ with components $\{dx^k/d\ell\}$ and the normal vector $\vec{N} = d\vec{T}/d\ell$ must satisfy the relations $\vec{T} \cdot \vec{T} = 1$ and $\vec{T} \cdot \vec{N} = 0$, which may be compared with (1.4.2) and (1.4.4).

The interest of these definitions resides in their intrinsic nature: the components U^α and Γ^α depend on the choice of frame, but the vectors $\vec{U}$ and $\vec{\Gamma}$ do not.

Chapter 2

Changes of reference system

In the preceding chapter we introduced the notion of a Minkowski frame. We shall introduce here the idea of an associated physical *reference system*.

Minkowski reference frames constitute a class of reference frames of which none is privileged, as the laws of physics are expressed in the same way in each one.

The coordinates of the same event in two different Minkowski reference frames are related by a *Poincaré transformation*. The invariance of physical laws under such a transformation constitutes a *principle of relativity*, to the extent that only the behaviour of a physical system *relative* to a reference frame can be observed.

These concepts are not specifically relativistic. We shall discuss them within the framework of both Newtonian and relativistic theories, so as to allow comparison and make precise the relation between the two theories.

2.1 Galilean transformations

In this section we consider only the Newtonian theory. The same questions are considered in the next section for the relativistic theory.

2.1.1 Free particle

We must now discuss the idea of a **free particle**. This concept completes classical kinematics and allows us to introduce the ideas of dynamics.

In Newtonian theory, a free particle is one whose velocity with respect to absolute space remains constant in time.

In practice it is not easy to decide if a particle is free. It is the coherence between this hypothesis and the set of physical results which attests its validity *a posteriori*. In the absolute Newtonian coordinate system we introduced earlier, the coordinates of a free particle obey the relations

$$x = V^1 t + x_0, \quad y = V^2 t + y_0, \quad z = V^3 t + z_0, \tag{2.1.1}$$

where V^k, x_0, y_0, z_0 are constants.

This is equivalent to

$$\blacklozenge \qquad \frac{d^2x}{dt^2} = \frac{d^2y}{dt^2} = \frac{d^2z}{dt^2} = 0.$$

However, there exist other Newtonian coordinate systems in which these relations hold if we change the numerical values of the constants. These coordinate systems are called **Galilean**.

2.1.2 The idea of a reference system

To construct a coordinate system, we first of all set in place a system of "measuring rods", allowing us to measure positions. At each point of space we place a clock.

> In reality we have only a finite supply of clocks, and we limit ourselves to measuring the positions of interesting objects, not necessarily by means of the measuring rods.

We have thus constructed a **reference system**, which allows us to find *by measurement* the coordinates *x*, *y*, *z*, *t* of events. A point mass, fixed by the measuring rods, will stay fixed with respect to this reference system. Its spatial coordinates *x*, *y*, *z* remain constant. We say that the coordinate system we have constructed is **comoving** with the matter (the measuring rods). A system of this type is not in general Galilean. However, imposing convenient conditions, we can construct a Galilean reference system to a certain approximation. To do this we must, for example, use "ideal" clocks, showing their own proper time, "synchronise" the time origin for each of them, use very rigid materials to construct the measuring rods, and maintain their orientations fixed with respect to the distant stars, etc. Each of these conditions poses a fundamental theoretical problem: the behaviour of clocks and materials, or the mysterious relation between the reference frame thus constructed and the distant stars, and so on.

Often objects are not situated close to a measuring rod. In order to use radar echos to determine the position of an aircraft and to compare this with the position which would be given by a measuring rod of the reference system, we have to know the laws governing the propagation of electromagnetic waves.

All of this is as complicated in practice as conceptually.

2.1.3 Galilean transformations

Let us assume that the reference system we have constructed is Galilean, or rather that within the accuracy of measurement there is no reason to assume otherwise.

Consider a second Galilean reference system in which an event has coordinates x', y', z', t': $x' = X'(x, y, z, t)$, $y' = Y'(x, y, z, t)$, $z' = Z'(x, y, z, t)$,

$t' = T'(x, y, z, t)$. To determine the functions X', Y', Z' and T' we must use the three relations (1.1.2), (1.1.3) and (2.1.1) which define Newtonian kinematics, and postulate their validity in the new reference system also.

The corresponding changes of coordinates are called **Galilean transformations**.

Among the various possible solutions, we describe as **special Galilean transformations** the following coordinate changes, denoted $G_x(V)$:

$$\blacklozenge \qquad\qquad t' = t, \quad x' = x - V t, \quad y' = y, \quad z' = z \qquad\qquad (2.1.2)$$

where V is a constant velocity.

To interpret these special Galilean transformations, consider the origin of the old frame, $x = y = z = 0$. In the new reference system this moves with velocity $-\vec{V}$ with components $\{-V, 0, 0\}$; $-\vec{V}$ is also the velocity with respect to the new reference system of all the points comoving with the old reference system: those for which x, y and z are constants. Similarly, considering the points comoving with the new reference system, we interpret $\vec{V} = \{V, 0, 0\}$ as the velocity of the new reference system with respect to the old.

We also define a **translation** of the coordinate system, with a change of time origin being regarded as a time translation:

$$x' = x + x_0, \quad y' = y + y_0, \quad z' = z + z_0, \quad t' = t + t_0,$$

with x_0, y_0, z_0, t_0 arbitrary constants.

The spatial reflection, R_x, corresponds to a change of orientation of the x axis:

$$x' = -x.$$

The most general Galilean transformation is the product of transformations $G_x(V^1)$, $G_y(V^2)$, $G_z(V^3)$ of an arbitrary translation, one or more reflections corresponding to one or more axes, and a global rotation of the axes with respect to the new origin.

We note that from the beginning, we have used the convention that a temporal event t_1 precedes an event t_2 iff $t_1 < t_2$. This convention could be altered and temporal reflections ($t' = -t$) considered as well; in this case we have $d\tau = dt = -dt'$.

If reflections are not considered, the most general transformation is determined by ten parameters: three parameters determine the special Galilean transformation, four parameters determine the most general translation, and three parameters determine a rotation (two angles define the axis of rotation and a third angle fixes the rotation around that axis). These ten parameters appear as constants of integration in the presentation below.

We now make precise the equations leading to the most general Galilean transformation.

Let A and B be specified by the coordinates $\{x, y, z, t\}$ and $\{x + dx, y + dy, z + dz, t + dt\}$ in the first Galilean coordinate system, and $\{x', y', z', t'\}$ and $\{x' + dx', y' + dy', z' + dz', t' + dt'\}$ in the second Galilean system.

The proper time of a body between two events is an intrinsic quantity, whose value is thus independent of the coordinate system. If both systems are Newtonian, the expression for $d\tau$ as a function of coordinates is formally the

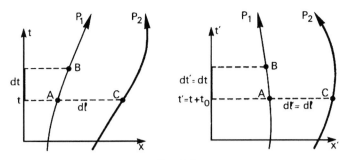

Fig. 2.1

same: $d\tau = dt = dt'$ for any dx, dy, dz (cf. (1.1.2) up to notation). We deduce that $t' = T'(x, y, z, t) = t + t_0$, where t_0 is an arbitrary constant; thus t and $t' = t + t_0$ represent the same instant in each of the two systems.

Consider a free particle whose motion is described by the relations (2.1.1) in the first coordinate system. We then demand that this motion be characterised by relations of the same type in the new system:

$$\frac{d^2x'(t')}{dt'^2} = \frac{d^2y'(t')}{dt'^2} = \frac{d^2z'(t')}{dt'^2} = 0$$

with $x' = X'(V^1t + x_0, V^2t + y_0, V^3t + z_0, t)$, $t = t' - t_0$ and similar expressions for y' and z', where V^k, x_0, y_0, z_0 are arbitrary constants.

This condition implies that X', Y', Z' are *linear* functions of x, y, z and t.

Consider two neighbouring particles P_1 and P_2 whose space coordinates at time t are $\{x, y, z\}$, $\{x + dx, y + dy, z + dz\}$ in the first coordinate system and $\{x', y', z'\}$, $\{x' + dx', y' + dy', z' + dz'\}$ in the second system (fig. 2.1):

$$x' = X'(x, y, z, t), \quad y' = Y'(x, y, z, t), \quad z' = Z'(x, y, z, t).$$

The spatial distance $d\ell$ separating the two bodies at the instant $t' = t + t_0$ is an intrinsic quantity. To determine it, we use (1.1.3) in the two coordinate systems:

$$d\ell^2 = (dx^2 + dy^2 + dz^2)_{(t)} = (dx'^2 + dy'^2 + dz'^2)_{(t'=t+t_0)}$$

with $(dx')_{t'} = \left(\dfrac{\partial X'}{\partial x}\right)_t dx + \left(\dfrac{\partial X'}{\partial y}\right)_t dy + \left(\dfrac{\partial X'}{\partial z}\right)_t dz$

and analogous expressions for $(dy')_{t'}$ and $(dz')_{t'}$.

Equating the coefficients of dx^2, dy^2, dz^2, $dx\,dy$, $dx\,dz$ and $dy\,dz$ on each side of this expression, we obtain six partial differential equations specifying the functions $X'(x, y, z, t)$, $Y'(x, y, z, t)$, $Z'(x, y, z, t)$.

The functions X', Y', Z' thus found, and the function T' found before, fix the Galilean transformations.

Consider a one-dimensional problem, i.e. assume that $t' = t + t_0$, $x' = X'(x, t)$, $y' = y$ and $z' = z$.

For a free particle we have

$$x' = X'(Vt + x_0, t) = V't' + x_0'; t' = t + t_0;$$

V, x_0, V', x_0', t_0 are constants; we get

$$\frac{d^2x'}{dt'^2} = \frac{d^2x'}{dt^2} = V^2\frac{\partial^2 X'}{\partial x^2} + 2V\frac{\partial^2 X'}{\partial x\,\partial t} + \frac{\partial^2 X'}{\partial t^2} = 0$$

for all V and x_0.

Taking the limit $V \to 0$, we deduce

$$\frac{\partial^2 X'}{\partial x^2} = \frac{\partial^2 X'}{\partial t^2} = \frac{\partial^2 X'}{\partial x\,\partial t} = 0 \quad \text{for } x' = X'(x_0, t) \text{ and any } x_0,$$

which implies:

$X' = ax + bt + c$ where a, b, and c are constants.

We thus get

$d\ell^2 = (dx')^2_{t'} = a^2 dx^2 = dx^2$ with $a = \pm 1$.

For $a = 1$, and $c = 0$, we find (2.1.2) in different notation.

2.2 Poincaré transformations

Here we study the same questions as in the previous section, but in special relativity.

2.2.1 The free particle

The notion of a free particle exists in relativity as a fundamental concept, but its definition is very different from that given in Newtonian theory. This is not surprising as the means of making it precise (space, velocity) have no intrinsic relativistic counterparts.

As space-time constitutes the only physical reality, we consider two events on the wordline of a particle.

How can we say if this worldline, $\mathscr{C}_0$, characterises the motion of a free particle?

To this end, we consider a second particle whose worldline, $\mathscr{C}$, joins the same two events. We calculate the proper time between the two events along $\mathscr{C}_0$ and along $\mathscr{C}$. Let τ_0 and τ be the respective values obtained.

The trajectory $\mathscr{C}_0$ is that of a free particle iff $\tau < \tau_0$, for all trajectories $\mathscr{C}$ (fig. 2.2).

Thus the free particle is that whose "ageing" between two given events is maximal. In appendix A we show that this condition makes the worldlines of free particles *in a system of Minkowski coordinates* take a form analogous to (2.1.1):

$$x = V^1 t + x_0, \quad y = V^2 t + y_0, \quad z = V^3 t + z_0,$$

or

$$\frac{d^2 x}{dt^2} = \frac{d^2 y}{dt^2} = \frac{d^2 z}{dt^2} = 0 \tag{2.2.1}$$

where V^k, x_0, y_0, z_0 are arbitrary constants subject to the condition $\sum_k (V^k)^2 < c^2$ (cf. § 1.2).

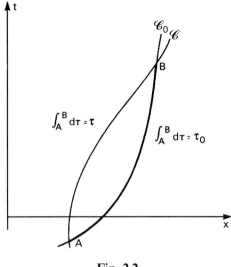

Fig. 2.2

We note that in Newtonian theory Newtonian reference systems are not necessarily Galilean, e.g. rotating frames. By contrast, in relativity the behaviour of a free particle is the same (eq. 2.2.1) in all Minkowski reference systems.

2.2.2 Minkowski reference systems

The accuracy with which a reference frame can be constructed, or recognised as Galilean, depends on the accuracy of the measurements involved. However, at the end of the 19th century, despite (because of) the advent of very accurate measurements, it proved impossible (in the Michelson–Morley experiment described in section 3.1) to realise in practice two Galilean reference systems related by a Galilean transformation. This difficulty was not a result of measurement errors, but had a disturbing systematic character.

We have already stressed that to construct a Galilean reference system one has to appeal to physical laws governing the systems involved, such as the rigidity of the materials used. The problem encountered in the Michelson–Morley experiment could have been a result of using the wrong physics, or have reflected instead defects in the Newtonian theory itself.

After many hesitations and discussions designed to "save" the Newtonian theory, the latter was abandoned in favour of relativity.

The procedure described above (§ 2.1.2) for constructing a reference system can obviously be retained, as the aim is simply to specify events. In relativity we need to know under what conditions the reference system is Minkowskian, i.e. when the resulting coordinates are Minkowski coordinates.

When all velocities are small compared with c Newtonian theory is extremely well verified. Special relativity is constructed so as to take account of this experimental fact. We have already shown how to do this for the kinematics.

For the laws of dynamics, we assume at the outset that Newton's laws hold in the limit of small velocities. We shall use this principle later to find the relativistic form of physical laws, which we then justify afterwards.

Under these conditions we can construct a "Galilean" reference system using Newtonian hindsight. Such a reference system is actually Minkowskian (we shall prove this in § 7.3.4). Henceforth we can pass from Newtonian to relativistic ideas just by interpreting observations in a relativistic fashion: that is, we change the set of interpretations without changing the descriptive framework.

2.2.3 Poincaré transformations

We assume that we have constructed a system of Minkowski coordinates x, y, z, t. Consider a second Minkowski system with coordinates x', y', z', t' expressed in terms of the old ones through functions X', Y', Z', T':

$$x' = X'(x, y, z, t), \quad y' = Y'(x, y, z, t), \quad z' = Z'(x, y, z, t), \quad t' = T'(x, y, z, t).$$

The proper time between two neighbouring events has an intrinsic value which is independent of the coordinate system. If both systems are Minkowskian we get from (1.2.1) that

$$c^2 d\tau^2 = c^2 dt^2 - dx^2 - dy^2 - dz^2 = c^2 dt'^2 - dx'^2 - dy'^2 - dz'^2. \qquad (2.2.2)$$

These relations are satisfied whatever the values of x, y, z, t, dx, dy, dz and dt (provided that $d\tau^2 > 0$); they can be interpreted as partial differential equations determining the functions X', Y', Z', T'. The corresponding changes of coordinate system are called **Poincaré transformations**.

To determine the Poincaré transformations we only need the form (1.2.1) of proper time while the three form-invariance conditions for time, space and motion of a free particle are needed to obtain the Galilean transformations. This simplification results from the fact that the expression (1.2.1) alone is the basis of relativistic kinematics.

Among all the possible solutions, we call the following, written $L_x(\beta)$, the **special Lorentz transformation**:

$$\begin{cases} x' = \dfrac{x - \beta ct}{\sqrt{1 - \beta^2}}, & y' = y, \quad z' = z \\[2mm] t' = \dfrac{t - \beta x/c}{\sqrt{1 - \beta^2}} \end{cases} \qquad (2.2.3)$$

We *interpret* the parameter β by proceeding in the same way as for the special Galilean transformation. Consider a point mass at rest with respect to the first reference system: thus x, y, z are constants. The worldline of this particle in the second reference system is naturally parametrised by t: $x' = -\beta\,c\,t/\sqrt{1-\beta^2}+\text{const}$, $y' = \text{const}'$, $z' = \text{const}''$, $t' = t/\sqrt{1-\beta^2}+\text{const}'''$. The "velocity" of this mass point with respect to the new reference frame has components

$$V'^x = \frac{dx'}{dt'} = -\beta c, \quad V'^y = \frac{dy'}{dt'} = 0 = V'^z = \frac{dz'}{dt'}.$$

Similarly, considering points at rest in the new reference system, we obtain the components of the "velocity" of the new reference system with respect to the old, expressed in the old frame: $V^x = \beta c$, $V^y = 0 = V^z$.

The most general Poincaré transformation has the same form as the Galilean transformation studied above. We have to substitute the product of transformations $L_x(\beta^1)$, $L_y(\beta^2)$, $L_z(\beta^3)$ for the product $G_x(V^1)$, $G_y(V^2)$, $G_z(V^3)$. Neglecting reflections, the Poincaré transformations depend on ten parameters also.

Consider a one-dimensional problem: $x' = X'(x, t)$, $y' = y$, $z' = z$ and $t' = T'(x, t)$. The equations satisfied by X' and T' are

$$c^2 d\tau^2 = c^2 dt^2 - dx^2 = c^2 \left\{ \frac{\partial T'}{\partial x} dx + \frac{\partial T'}{\partial t} dt \right\}^2 - \left\{ \frac{\partial X'}{\partial x} dx + \frac{\partial X'}{\partial t} dt \right\}^2$$

or

$$c^2 \left(\frac{\partial T'}{\partial t} \right)^2 - \left(\frac{\partial X'}{\partial t} \right)^2 = c^2, \quad c^2 \left(\frac{\partial T'}{\partial x} \right)^2 - \left(\frac{\partial X'}{\partial x} \right)^2 = -1,$$

$$c^2 \left(\frac{\partial T'}{\partial x} \right) \left(\frac{\partial T'}{\partial t} \right) - \left(\frac{\partial X'}{\partial x} \right) \left(\frac{\partial X'}{\partial t} \right) = 0.$$

The first two equations can be written as

$$\frac{\partial T'}{\partial t} = \varepsilon_1 \cosh\phi, \quad \left(\frac{1}{c} \right) \frac{\partial X'}{\partial t} = \sinh\phi, \quad c\frac{\partial T'}{\partial x} = \sinh\Theta, \quad \frac{\partial X'}{\partial x} = \varepsilon_2 \cosh\Theta$$

with $(\varepsilon_1)^2 = (\varepsilon_2)^2 = 1$. The third relation leads to the equality $\varepsilon_2\phi = \varepsilon_1\Theta$. We thus show that the system is equivalent to

$$\frac{\partial T'}{\partial t} = \varepsilon\frac{\partial X'}{\partial x}, \quad \frac{\partial T'}{\partial x} = \varepsilon \left(\frac{1}{c^2} \right) \frac{\partial X'}{\partial t}, \quad \left(\frac{\partial T'}{\partial t} \right)^2 - c^2 \left(\frac{\partial T'}{\partial x} \right)^2 = 1$$

with $\varepsilon = \varepsilon_2/\varepsilon_1 = \pm 1$.

Eliminating $\partial^2 X'/\partial t\,\partial x$ from the first two equations we get an equation for T':

$$\left(\frac{1}{c^2} \frac{\partial^2 T'}{\partial t^2} \right) - \left(\frac{\partial^2 T'}{\partial x^2} \right) = 0.$$

The general solution of this equation is $T' = F(t - x/c) + G(t + x/c)$, where F and G are arbitrary functions. Substituting this expression for T' in the last equation gives

$$4 \left(\frac{\partial F}{\partial t} \right) \left(\frac{\partial G}{\partial t} \right) = 1.$$

Considering the values $t = \pm x/c$, we can show that $dF(u)/du = $ const and $dG(u)/du = $ const'. The two constants are not independent: $dF/du = 1/(2a)$ and $dG/du = a/2$, where a is an arbitrary constant. Integrating and substituting the result in the original equations we get on integration the general solution in the form

$$X'(x, t) = \varepsilon \left\{ \left(\frac{a}{2} - \frac{1}{2a} \right) ct + \left(\frac{a}{2} + \frac{1}{2a} \right) x \right\} + x_0$$

$$T'(x, t) = \left\{ \left(\frac{a}{2} + \frac{1}{2a} \right) t + \left(\frac{a}{2} - \frac{1}{2a} \right) \frac{x}{c} \right\} + t_0$$

where a, x_0, t_0 are arbitrary constants and $\varepsilon = \pm 1$.

These expressions can also be written as

$$x' = \varepsilon_1 \frac{x - (\beta ct)}{\sqrt{1 - \beta^2}} + x_0 \quad t' = \varepsilon_2 \frac{t - \left(\beta \frac{x}{c} \right)}{\sqrt{1 - \beta^2}} + t_0$$

where β is an arbitrary constant with modulus less than or equal to unity, $\beta = (1 - a^2)/(1 + a^2)$, $|\beta| \leqslant 1$. For $\varepsilon_1 = \varepsilon_2 = 1$, $x_0 = t_0 = 0$, we find (2.2.3).

2.3 Change of basis of T_4 and addition of velocities

2.3.1 Change of basis of T_4

We have seen that a system of Minkowski coordinates of M_4 allowing us to specify events is associated with a natural basis of T_4, written $\{ \vec{e}_\alpha \}$. We now know how the coordinates transform when we pass from a Minkowski system $\{ x^\alpha \}$ to another Minkowski system $\{ x'^\alpha \}$. We now wish to determine the new natural basis $\{ \vec{e}'_\alpha \}$ of T_4 associated with the new Minkowski coordinates. We restrict ourselves to studying special Lorentz transformations.

The event at the origin of M_4, Ω, is the same in both reference systems: $x_\Omega = y_\Omega = z_\Omega = t_\Omega = 0 = x'_\Omega = y'_\Omega = z'_\Omega = t'_\Omega$ (cf. 2.2.3).

Let E be an arbitrary event of M_4, and $\overrightarrow{\Omega E}$ the corresponding vector of T_4:

$$\overrightarrow{\Omega E} = \sum_\alpha x^\alpha \cdot \vec{e}_\alpha = \sum_\alpha x'^\alpha \cdot \vec{e}'_\alpha.$$

This relation defines $\{ \vec{e}'_\alpha \}$. We set

$$x = x^1, \quad x' = x'^1, \quad y = x^2, \quad y' = x'^2, \quad z = x^3, \quad z' = x'^3$$
$$ct = x^0, \quad ct' = x'^0.$$

Using (2.2.3) we find

$$\vec{e}'_1 = \frac{\vec{e}_1 - \beta \cdot \vec{e}_0}{\sqrt{1 - \beta^2}} \qquad \vec{e}'_0 = \frac{\vec{e}_0 - \beta \cdot \vec{e}_1}{\sqrt{1 - \beta^2}} \tag{2.3.1}$$

$$\vec{e}'_2 = \vec{e}_2 \qquad \vec{e}'_3 = \vec{e}_3. \tag{2.3.2}$$

Using the formulae for a change of basis we immediately get the components of the four-velocity in the new basis:

$$\begin{cases} \vec{U} = \sum_\alpha U^\alpha \cdot \vec{e}_\alpha = \sum_\alpha U^{\alpha'} \cdot \vec{e}'_\alpha; \\[2mm] U^{0'} = \dfrac{U^0 - \beta U^1}{\sqrt{1-\beta^2}}; \quad U^{1'} = \dfrac{U^1 - \beta U^0}{\sqrt{1-\beta^2}}; \\[2mm] U^{2'} = U^2 \quad \text{and} \quad U^{3'} = U^3. \end{cases} \qquad (2.3.3)$$

The method used to find $U^{\alpha'}$ relies on the fact that $\vec{U}$ is a four-vector. We thus verify the relation $U^{\alpha'} = dx'^\alpha/d\tau$. Using (2.2.3) and (2.2.2), we could have found $U^{\alpha'} = dx'^\alpha/d\tau$ and verified the relations (2.3.3). The latter guarantee that U^α and $U^{\alpha'}$ are the components of the same four-vector.

2.3.2 Addition of velocities

Consider a body M moving with respect to a Minkowski reference system. We can specify its "position" at any "instant". In reality we only determine the time at which the measuring apparatus detects the presence of the body. We can thus parametrise the worldline as in Newtonian theory: $x^k = x^k(t)$. We introduce a matrix notation in which $\overline{V}$ represents the matrix with components V^k and $\overline{U}$ the matrix with components U^k.

$$V^k = \frac{dx^k}{dt} \qquad U^k = \frac{dx^k}{d\tau}$$

with $d\tau = \sqrt{(1-\beta^2)}\, dt$ and $\beta^2 = (\sum_k V^k \cdot V^k)/c^2 = \overline{V} \cdot \overline{V}/c^2$. This notation allows us formally to use three-dimensional vector algebra (We will drop this notation in part II and use the usual vector notation $\vec{V}$ rather than $\overline{V}$). We interpret $\overline{V}$ as the "velocity" of the body with respect to the reference system under consideration; $\overline{V}$ is actually measurable. U^α are then explicitly determined as functions of measured quantities:

$$U^0 = \frac{cdt}{d\tau} = \frac{c}{\sqrt{1-\beta^2}}, \quad U^k = \frac{dx^k}{d\tau} = \frac{V^k}{\sqrt{1-\beta^2}}. \qquad (2.3.4)$$

We now consider two Minkowski reference frames R and R'. Let $\overline{V}_0$ be the "velocity" of R' with respect to R. Let a body have "velocity" $\overline{V}$ with respect to the first reference system and $\overline{V}'$ with respect to the second. The problem of adding velocity reduces to determining $\overline{V}'$ as a function of $\overline{V}$ and $\overline{V}_0$. We restrict ourselves to the case where R and R' are related by a special Lorentz transformation (2.2.3) with $\overline{V}_0 = \{V_0, 0, 0\}$ and $\beta_0 = V_0/c$.

From (2.3.4) we get

$$V^k = c\frac{U^k}{U^0} \qquad V^{k'} = c\frac{U^{k'}}{U^{0'}}.$$

Using the relations (2.3.3) we get

$$V^{1'} = c\frac{U^1 - (\beta_0 U^0)}{U^0 - (\beta_0 U^1)} = \frac{V^1 - V_0}{1 - \left(V_0\dfrac{V^1}{c^2}\right)}$$

$$V^{2'} = cU^2\frac{\sqrt{1 - \beta_0^2}}{U^0 - (\beta_0 U^1)} = \frac{V^2\sqrt{1 - \beta_0^2}}{1 - \left(V_0\dfrac{V^1}{c^2}\right)}$$

$$V^{3'} = cU^3\frac{\sqrt{1 - \beta_0^2}}{U^0 - (\beta_0 U^1)} = \frac{V^3\sqrt{1 - \beta_0^2}}{1 - \left(V_0\dfrac{V^1}{c^2}\right)}.$$

We obtain the low-velocity limit by setting $V_0/c = 0$ or letting c tend to infinity. In this case we recover the Newtonian velocity addition law.

For $\parallel \overline{V} \parallel = c$ we find $\parallel \overline{V'} \parallel = c$: the fundamental velocity is the same in every Minkowski reference frame. We can also deduce this from the relation $d\tau = 0$ characterising motions of velocity c.

2.4 Relativity principles

2.4.1 The Galilean relativity principle

We have used two important ideas to determine Galilean coordinate systems, and we first reemphasise and then extend them here.

(a) The Galilean transformations we have considered up to now do not deal with the physical systems but only with their description. They are thus called **passive transformations**. It is clear that physical predictions cannot depend on the mode of description used. Thus the numerical value of proper time between two events does not depend on the reference system used.

(b) Galilean transformations allow us to show the existence of a set of privileged reference systems characterised by
– the expression (1.1.2) for proper time,
– the expression (1.1.3) for the distance between two bodies, and
– the expression (2.1.1) for the motion of a free particle.

These expressions are *form invariant* under Galilean transformations; this means for example that the expression for the proper time in terms of the coordinates remains the same in any Galilean system.

(c) Now consider the following problem (fig. 2.3):

In a Galilean reference system R, the motion of a body M is characterised by $X = F(t)$, $Y = G(t)$, $Z = H(t)$. The events E_1 and E_2 have coordinates $x_1 = F(t_1)$, $y_1 = G(t_1)$, $z_1 = H(t_1)$, $t = t_1$ and $x_2 = F(t_2)$, $y_2 = G(t_2)$, $z_2 = H(t_2)$, $t = t_2$.

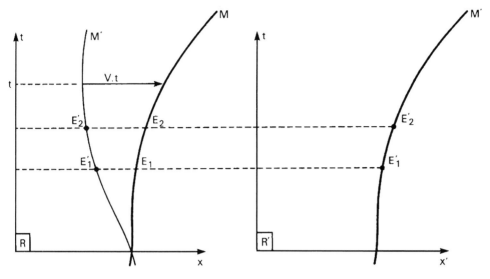

Fig. 2.3

Let us consider a second body, M', whose coordinates in the same reference system are deduced from the first by a special Galilean transformation: $X' = X - Vt = F(t) - Vt$, $Y' = Y$, $Z' = Z$. The events E_1' and E_2' correspond to E_1 and E_2 under the same transformation; they have coordinates

$$x_1' = x_1 - Vt_1, \ y_1' = y_1, \ z_1' = z_1, \ t_1' = t_1$$
$$\text{and } x_2' = x_2 - Vt_2, \ y_2' = y_2, \ z_2' = z_2, \ t_2' = t_2.$$

Here we have used the Galilean transformation to create a new physical situation and not for describing the same situation in a different way. This kind of transformation is called an **active transformation**.

Let us apply the same active transformation to the set of point masses constituting the reference system R, and assume that the clocks whose motions have been changed continue to indicate their proper time: we obtain a reference system R'. The coordinates of an event in R' are deduced from those of the same event in R by the Galilean transformation $t' = t$, $x' = x + Vt$, $y' = y$, $z' = z$. The reference system R' is thus Galilean.

The motion of M' is described in R' in an identical way to that of M in R, up to the notation.

We easily see that the proper time, e.g. between E_1 and E_2, is the same as the proper time between $E_{1'}$ and $E_{2'}$.

This is a consequence of the form-invariance of the expression for proper time.

The two physical situations, M with respect to R and M' with respect to R', are indistinguishable by measurements of proper time, and of any quantity form-invariant under passive Galilean transformations.

More generally, as the laws of Newtonian mechanics are form-invariant under arbitrary Galilean transformations, it is impossible to pick out either of the two reference systems $\{M, R\}$, $\{M', R'\}$ by experiments involving mechanics alone. More concisely, as the Galilean transformations form a group, we say that the Galilean group is the *symmetry group of Newtonian mechanics*. This assertion, raised to the level of a principle, constitutes the **Galilean relativity principle**. According to this principle, no Galilean reference system can be recognised as privileged as a result of mechanical experiments; absolute physical motions cannot be observed, as only physical processes *relative* to a reference system are accessible to experiment.

The question arises of how one can perform the most general active Galilean transformation to pass from R to R' in practice. The simplest way is to give R a motion to reorient it and give it a suitable velocity. This procedure is conceptually acceptable in the limit where accelerations are infinitesimal, if we assume that the effects of the motion imposed occur only through inertial forces. As the inertial forces are negligible the matter remains undistorted; this is meaningful in Newtonian theory, where lengths are *a priori* quantities. Moreover, the internal working of the clocks is unchanged by the presence of infinitesimal inertial forces: they continue to show their proper time.

Newton's theory leads to the impossibility of detecting absolute motion through mechanical experiments. However, from the beginning of the 19th century physicists thought that optical experiments might allow one to detect such motions: for example the motion of the Earth in its orbit around the Sun. This attempt failed.

2.4.2 The wave equation

Consider a wave phenomenon propagating with speed c_0 in a Galilean reference system R. The corresponding physical quantity is written ϕ. A set of observers is supposed to be able to measure ϕ; these measurements constitute events with coordinates x, y, z, t. The corresponding values of ϕ define the function $\phi(x, y, z, t)$. We restrict ourselves to studying a one-dimensional problem where ϕ is independent of y and z: $\phi = \phi(x, t)$.

The wave equation satisfied by ϕ is

$$\frac{\partial^2 \phi}{c_0^2 \partial t^2} - \frac{\partial^2 \phi}{\partial x^2} = 0. \tag{2.4.1}$$

This equation has the general solution $\phi = F(t - x/c_0) + G(t + x/c_0)$ where F and G are arbitrary functions. These functions represent waves moving in opposite directions with speed c_0.

Let us make a Galilean transformation $x' = x - Vt$, $t' = t$ and set $\Phi(x', t') = \phi(x, t)$. The function Φ describes the wave in a new Galilean reference system R'. We seek the wave equation satisfied by Φ:

$$\frac{\partial \phi}{\partial t} = -\left(V \frac{\partial \Phi}{\partial x'} \right) + \frac{\partial \Phi}{\partial t'}, \quad \frac{\partial \phi}{\partial x} = \frac{\partial \Phi}{\partial x'}$$

$$\frac{\partial^2 \phi}{\partial t^2} = \frac{\partial^2 \Phi}{\partial t'^2} - \left(2V \frac{\partial^2 \Phi}{\partial x' \partial t'} \right) + \left(V^2 \frac{\partial^2 \Phi}{\partial x'^2} \right), \quad \frac{\partial^2 \phi}{\partial x^2} = \frac{\partial^2 \Phi}{\partial x'^2}.$$

Using (2.4.1) we get

$$\frac{1}{c_0^2} \left\{ \frac{\partial^2 \Phi}{\partial t'^2} - \left(2V \frac{\partial^2 \Phi}{\partial x' \partial t'} \right) + \left(V^2 \frac{\partial^2 \Phi}{\partial x'^2} \right) \right\} - \frac{\partial^2 \Phi}{\partial x'^2} = 0. \qquad (2.4.2)$$

The wave equation has changed form: it is not invariant under Galilean transformations. This lack of invariance allows us to measure V, by determining the coefficients of the wave equation in the new reference system.

Maxwell's theory of electromagnetism leads to equations of the form (2.4.1) above for the components of the electric field associated with a wave propagating in a vacuum; c_0 is thus the speed of light. For the whole of the 19th century, physicists thought that light was carried by an all-pervasive medium, the ether. It was reasonable to assume that equation (2.4.1) was satisfied by light waves in a reference system tied to the ether, while (2.4.2) was satisfied in a terrestrial reference system moving with respect to the ether.

Actually attempts to measure the speed of the Earth through the ether began more than half a century before Maxwell. The motives for this research were not formulated in terms of "lack of invariance of the wave equation under the Galilean group", but in a way that was, however, equivalent.

All attempts to measure the velocity V of the Earth with respect to the ether were doomed to failure, including that of Michelson and Morley (1887) which we discuss below.

2.4.3 Identification of the fundamental velocity c with the velocity of light in vacuo

The problems connected with the asymmetric treatment of electric and magnetic fields in two Galilean reference systems led Einstein to special relativity in 1905.

Special relativity did not arise directly from attempts to explain the Michelson–Morley experiment; however, this experiment is very important, in suggesting that the velocity of light is the same in all Galilean reference systems. We shall show how this result is interpreted in relativity. We shall consider this experiment from another point of view in the following chapter.

Let us assume that in a Minkowski reference system R, the wave equation has the form (2.4.1). Following the same method as above, we find the wave equation satisfied by Φ in a new Minkowski reference system R'. The only modification comes from the change of coordinates, which now has the form (2.2.3):

$$x' = \frac{x - \beta ct}{\sqrt{1 - \beta^2}}, \quad t' = \frac{t - \beta \dfrac{x}{c}}{\sqrt{1 - \beta^2}},$$

We thus get

$$\left[\left(\frac{1}{c_0^2} - \frac{\beta^2}{c^2}\right)\frac{\partial^2\Phi}{\partial t'^2}\right] - \left[2\beta\left(\frac{c}{c_0^2} - \frac{1}{c}\right)\frac{\partial^2\Phi}{\partial x'\partial t'}\right] - \left[\left(1 - \frac{\beta^2 c^2}{c_0^2}\right)\frac{\partial^2\Phi}{\partial x'^2}\right] = 0.$$

(2.4.3)

The coefficients in this equation depend on the velocity $c\beta$ of the reference system R' with respect to R. We can in general find this velocity experimentally starting from the wave equation. However, for $c = c_0$, equations (2.4.1) and (2.4.3) have the same form. In this case it is impossible to distinguish the reference systems R and R' and to determine β.

To interpret the experimental results, particularly the results of the Michelson–Morley experiment, we identify the fundamental quantity c, introduced in (1.2.1), and the speed c_0 of electromagnetic waves in vacuo. The wave equation for light then conserves its form in all Minkowski reference systems, and the speed of light waves retains the same value.

We can thus see why Newtonian theory seemed so well verified. The reference systems which were thought to be Galilean were in reality Minkowski reference systems in special relativity. As long as the low-velocity approximation holds the numerical predictions of the two theories are identical, even if their interpretations differ. When experiments involve light waves, the low-velocity approximation fails, as the velocity of the photons is precisely the limiting velocity of Einstein's theory.

2.4.4 Einstein's principle of relativity

The analysis we have given for Galilean transformations can be repeated for Poincaré transformations.

(a) Passive Poincaré transformations allow several descriptions of the same physical reality in different Minkowski reference systems. The predictions do not depend on the description adopted.

(b) Poincaré transformations are those transformations preserving the form invariance of the expression (1.2.1) for proper time. This condition alone fixes them. In this respect the situation is simpler than in Newtonian theory.

(c) We adopt the idea of form invariance of all physical laws under passive Poincaré transformations as a principle. According to this principle no Minkowski reference system is privileged over any other. Only physical processes with respect to a reference system are accessible to experiment and in this sense all Minkowski reference systems are equivalent. This constitutes **Einstein's principle of relativity.**

To carry out an active Poincaré transformation and move from one Minkowski reference system R to another, R', we proceed as described for Galilean transformations. However, in relativity we cannot claim that the matter remains undeformed, as this kind of statement refers to an absolute form, which has no physical meaning *a priori*; only the world-tube has physical reality.

Chapter 3

Experiments

We leave the detailed study of the experiments discussed in this chapter as an exercise for the reader.

Let us consider a theory furnished with a descriptive framework and a body of interpretations. To "test" this theory, we carry out a number of experiments and compare the results with the predictions of the theory. If the differences are smaller than the measurement uncertainties we say that the theory is "verified". In reality other, "neighbouring" theories, which may be unknown, are as well "verified". If a theory disagrees with the results of experiment, it is not easy to decide if the theory is to be rejected, or that the disagreement signals a new effect within the same theory. Thus no "crucial" experiments exist in an absolute sense. However, in a given set of theories, for example in comparing two theories, crucial experiments can exist.

In this chapter we shall discuss several experiments, with the aim of comparing the Newtonian and relativistic predictions. We shall see that relativity theory is in agreement with observation to within experimental error. Newtonian theory might perhaps be saved, but at the price of introducing a number of effects designed with this sole end in view, which we would then have to explain.

The descriptive framework of our discussions is common to both theories. In particular we assume that Galilean reference systems in Newtonian theory, viewed in the low-velocity limit, are in reality the Minkowski reference systems of relativity.

In this framework there are two possible interpretations for the same phenomenon; the differences between the predictions of the Newtonian and relativistic theories become observable when the velocities are no longer small. We can thus choose the "better" theory.

We assert that the "better" theory is that in which there is no need to introduce *ad hoc* effects to account for the differences between prediction and observation. This criterion involves essentially an "aesthetic" choice, rather than absolute necessity. The history of physics shows how fruitful this type of choice is.

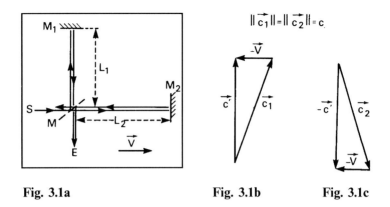

Fig. 3.1a   Fig. 3.1b   Fig. 3.1c

3.1 The Michelson–Morley experiment

We analyse this experiment in a Newtonian framework.

Consider an optical experiment carried out in a laboratory moving with velocity $\vec{V}$ with respect to the ether. We set $V = \|\vec{V}\|$. Let G be the measured quantity. We can expand G in a power series of the form $G = G_0 + \beta G_1 + \beta^2 G_2 + \ldots$, with $\beta = V/c$, where c is the speed of light with respect to the ether. The term βG_1 is called the **first order term**, and $\beta^2 G_2$ the **second order term**. The laboratory is rigidly fixed to the Earth, whose velocity in its orbit about the Sun is $V_0 \approx 30$ km s^{-1}. We assume for simplicity that the Copernican frame centred on the Sun, with axes pointing to the distant stars, is stationary with respect to the ether. We can thus take $\beta \approx 10^{-4}$. Second order effects are difficult to detect, so physicists at first put their efforts into demonstrating first order effects.

The first experiments by Arago in 1818 proved negative: no first order effect was observed ($G_1 = 0$). In the same year Fresnel suggested that the ether might be dragged along with velocity $\vec{U} = (1 - 1/n^2)\vec{V}$ depending on the refractive index n of the medium through which the light propagated. All first order experiments were negative, thus confirming the validity of Fresnel's formula in each case. In 1874, Mascart, Veltmann and Pottier showed that this formula wrecked any possibility of observing a first-order effect in optics.

In 1881, Michelson performed an experiment designed to measure a second order effect. The results he announced were again negative: $G_2 = 0$. However, an error in calculation had led Michelson to overestimate the expected effect. As the real theoretical effect was smaller, it could have been masked by measurement errors.

In 1887, Michelson and Morley repeated the 1881 experiment with increased sensitivity. The result was still negative.

To measure the effect being sought, Michelson used an interferometer of the type shown schematically in fig. 3.1a in both experiments.

The light emitted by the source S is divided into two beams by a splitter M, which is a semi-transparent mirror. The two beams interfere at E after reflection at the mirrors M_1 and M_2 (fig. 3.1a). The arms have lengths L_1 and L_2. We assume that the whole apparatus moves with velocity $\vec{V}$ with respect to the ether, parallel to MM_2.

The velocity of light with respect to the laboratory is $c - V$ from M to M_2 and $c + V$ from M_2 to M. The time taken by light over the path MM_2M is t_2:

$$t_2 = \frac{L_2}{c - V} + \frac{L_2}{c + V} \approx 2L_2 \, \frac{1 + \beta^2}{c}.$$

From M to M_1 the velocity c' of light with respect to the laboratory can be found from the law of addition of velocities, as shown on fig. 3.1b, where $\vec{c}_1$ is the velocity of light with respect to the ether; $\|\vec{c}_1\| = c$. We find $c' = \|\vec{c}'\| = \sqrt{(c^2 - V^2)}$. This velocity is also the velocity of the light from M_1 to M (fig. 3.1c). The time t_1 along the path MM_1M is thus

$$t_1 = \frac{2L_1}{\sqrt{c^2 - V^2}} \approx 2L_1 \, \frac{1 + \dfrac{\beta^2}{2}}{c}.$$

In 1881, Michelson assumed t_1 independent of β: this is the error mentioned above.

As the waves are in phase at S, they must be out of phase at E by $\delta\Phi = 2\pi(t_1 - t_2)/T$ where T is the period of the light waves. Using the wavelength $\lambda = cT$, we find

$$\delta\Phi = 4\pi \, \frac{L_1 - L_2}{\lambda} + 2\pi \, \frac{L_1 - 2L_2}{\lambda} \, \beta^2.$$

Let us rotate the apparatus through 90° in its plane. An identical calculation gives the phase difference

$$\delta\Phi' = 4\pi \, \frac{L_1 - L_2}{\lambda} + 2\pi \, \frac{2L_1 - L_2}{\lambda} \, \beta^2.$$

Thus in the rotation, the phase difference changes by

$$\delta\Phi' - \delta\Phi \approx 2\pi \, \frac{L_1 + L_2}{\lambda} \, \beta^2.$$

By using multiple reflections in each arm, Michelson and Morley obtained lengths $L_1 \approx L_2 \approx 11$ m.

With $L_1 + L_2 \approx 20$ m and $\lambda \approx 0.5$ μm, we find a phase change corresponding to 0.4 fringe spacings for $\beta^2 \approx 10^{-8}$.

The observed phase changes were less than 0.02 fringe spacings, and were entirely attributable to experimental error.

We now know the explanation for this negative result: the arms of the interferometer, before and after rotation, are not two Galilean reference systems of Newtonian mechanics, but two Minkowski reference systems of relativity. The Newtonian law of addition of velocities cannot be applied: the "velocity" of light with respect to each system (cf. § 2.3) is c in all directions.

Before everything was explained within the framework of relativity, Fitzgerald and Lorentz suggested the existence of a new Newtonian effect. The law of addition of velocities would remain correct but lengths would be changed in the rotation.

The arm MM_1 would contract when aligned along $\vec{V}$; its length would become $L_1\sqrt{(1-\beta^2)}$, while the arm MM_2 orthogonal to $\vec{V}$ would dilate. Thus a ruler moving with respect to the ether would contract, but we could not check this with another ruler as the second ruler would have contracted in the same way.

To explain the lack of a first order effect, one had to assume that the ether was dragged along as specified by Fresnel's formula. To explain the absence of a second order effect, one had to assume the Fitzgerald–Lorentz contraction. Both of these effects themselves required explanation: by contrast relativity had the supreme merit of economy.

The comparison of Newtonian and relativistic interpretations shows that relativity is the "better" of the two theories, provided the fundamental velocity of (1.2.1) is identified with the velocity of light in vacuo (cf. § 2.4.3).

We thus accept the body of relativistic interpretations and assume furthermore that the reference system formed by the arms of the interferometer in their first position is Minkowskian. The Michelson–Morley experiment then shows that the arms suffer an active Poincaré transformation in the rotation.

Nowadays the confidence in relativity is such that the velocity of light is chosen as a fundamental unit along with the second; the metre is now a derived unit.

> The Michelson–Morley experiment thus shows that in the rotation the length difference between the two arms is less than $0.5 \times 10^{-9}L$. A modern version of this experiment was performed by A. Brillet and J. Hall in 1977 at Boulder in the United States. It confirmed the results of Michelson and Morley with an accuracy of at least 10^{-12} and perhaps of order 10^{-15} (these numbers are comparable with the 0.5×10^{-9} of the Michelson–Morley experiment, although the experiments were quite different).

3.2 Time

In kinematics, the main effect of relativity is the interpretation of time, which loses its absolute character. We discuss some aspects here.

3.2.1 The Doppler effect

The study of spectral shifts and their interpretation is very important, particularly in astrophysics and cosmology.

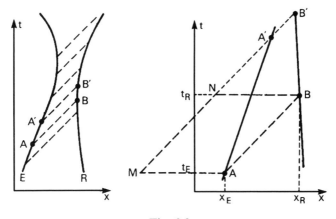

Fig. 3.2

The problem is as follows. An emitter E sends a periodic light signal to a receiver R. The emitter measures an emission period T_E while the receiver receives a signal of period T_R. During the emission the velocity of the emitter is constant; that of the receiver remains constant over the time of reception. This is usually true for observations of electromagnetic waves in particular. These have high frequencies, and the velocities of the emitter and receiver do not vary significantly over the time required to measure them. We stress that we do not assume that the velocities remain constant over the light travel time. In fact it takes several years for light to reach us from the nearest star, and over this time the Earth's velocity in its orbit around the Sun for example has not remained constant.

> The last paragraph used Newtonian language; as an exercise, the reader should pick out the Newtonian ideas and make their relativistic meanings precise.

To study this problem and compare the relativistic and Newtonian interpretations we consider the simple case where all the motions are in the $\{x, t\}$ plane.

In a Minkowski reference system, the emitter has velocity $\overline{V}_E = \{V_E, 0, 0\}$ at the moment of emission, while the receiver has velocity $\overline{V}_R = \{V_R, 0, 0\}$ at the moment of reception.

As very short light signals are emitted regularly, they can be regarded as points of light with coordinates $X_L = c(t + \text{const})$, $Y_L = 0$, $Z_L = 0$.

Fig. 3.2 shows the various motions.

Event A, $\{x_E, t_E\}$, corresponds to the emission of the first signal. Its reception is the event B, $\{x_R, t_R\}$. The emission A', and reception B', of the following signal have coordinates $\{x'_E, t'_E\}$ and $\{x'_R, t'_R\}$ respectively.

In fig. 3.2 we have $BN = AM$, or

$$(c - V_R)(t'_R - t_R) = (c - V_E)(t'_E - t_E). \tag{3.2.1}$$

The period at emission, T_E, is the proper time of the emitter between A and A'. The period at reception, T_R, is the proper time of the receiver between B and B'. Using the expression (1.2.2) for proper time we get

$$T_E = \sqrt{1 - \frac{V_E^2}{c^2}}\,(t'_E - t_E)$$

$$T_R = \sqrt{1 - \frac{V_R^2}{c^2}}\,(t'_R - t_R).$$

Using (3.2.1), we get

$$\frac{T_R}{T_E} = \sqrt{\frac{1 + \dfrac{V_R}{c}}{1 - \dfrac{V_R}{c}}} \cdot \sqrt{\frac{1 - \dfrac{V_E}{c}}{1 + \dfrac{V_E}{c}}}. \tag{3.2.2}$$

This kinematic effect, in which T_E differs from T_R, is *the Doppler effect*.

When we observe a spectral line from a distant atom, its period is generally not the same as the period of the same line emitted by the same atom on the Earth. This is called the **spectral shift**. The Doppler effect is one of the causes of the observed spectral shifts.

In the low-velocity approximation we get

$$\frac{T_R}{T_E} = 1 + \frac{V_R - V_E}{c}. \tag{3.2.3}$$

Interpreting the phenomenon in the framework of Newtonian theory, we would have regarded the reference system as Galilean and c as the velocity of the signal in this system. The mathematical relation (3.2.1) remains unchanged. The expressions for the proper times are in contrast changed: we must use (1.1.2). We then find:

$$T_E = (t'_E - t_E) \quad \text{and} \quad T_R = (t'_R - t_R)$$

$$\frac{T_R}{T_E} = \frac{1 - \dfrac{V_E}{c}}{1 - \dfrac{V_R}{c}}.$$

In the limit of small velocities this expression reduces to (3.2.3), which holds under the same conditions.

Many experiments confirm the validity of (3.2.2). The Doppler effect is thus naturally interpreted in relativity.

3.2.2 The twin paradox

A pair of twins are separated at birth. One stays on the Earth, while the other leaves in a rocket on a long journey (long by which clock?). The rocket's velocity stays constant except for short intervals during which it is accelerated to take off and turn round. When the rocket returns to Earth the twin from the rocket is still young but his brother is very old.

This relativistic fable caused considerable shock as it undermined the dogma of universal time planted in us very early, through the use of Newtonian language, and because it appeared to cast doubt on the principle of relativity.

To pose the problem precisely, we must assume that the Earth twin remains at rest with respect to a Minkowski reference system. The other twin undergoes accelerations, and, however short these are, "knows" that his rest-frame is not Minkowskian. Accelerometers attached to each of the twins behave differently, and no appeal to any symmetry principle is possible, still less a relativity principle: the problem involves more than the relative motion of the two twins. The twin attached to a Minkowski reference system moves like a free particle, and therefore "ages" more than his brother (cf. § 2.2).

But we should ask what experiment tells us here.

Biological systems are not precise enough clocks to be used to verify relativity at present. Since the frequency stability of an atomic clock greatly exceeds that of a human heart, however, it has been possible to confirm the twin "paradox" using such clocks. We describe here an idealised experiment based on the same principles as those used in practice.

Two aeroplanes circle the Earth in the equatorial plane in opposite directions at the same constant altitude, with velocity $V \approx 1000$ km h^{-1}. We consider a reference frame centred on the Earth, whose axes always point in the same directions with respect to the fixed stars. We assume that this reference system is Minkowskian, as can be justified in several ways (cf. § 2.2.2). In this reference frame, fig. 3.3 shows the helices described by the worldline of each aeroplane in a space–time diagram.

The Earth rotates with uniform angular velocity $\Omega \approx 7.3 \times 10^{-5}$ rad s^{-1}. The aeroplanes describe circles in the $z = 0$ plane. The radius R of this circle is of order $R \approx 6400$ km. The velocities of the aeroplanes with respect to the reference system are thus $V - R\Omega$ and $V + R\Omega$ respectively. Since all the velocities are small compared with c, we can use Newtonian arguments. The expected effect is small, and a detailed relativistic analysis of the velocity measurements only gives a very small change to a result which is itself already small. There is thus no point in such an analysis here.

When the aeroplanes pass each other at times t and t', after completing a circle, clocks on board each of them are compared. The proper time of each clock is given by (1.2.2):

$$\tau_+ = \int_t^{t'} \sqrt{1 - \beta_+^2}\, dt \qquad \tau_- = \int_t^{t'} \sqrt{1 - \beta_-^2}\, dt$$

where $\beta_+ = (V + R\Omega)/c$, $\beta_- = (V - R\Omega)/c$ and $t' - t = 2\pi R/V$. Thus the two clocks have aged differently: $(\tau_-) - (\tau_+) \approx (R\Omega/c) \cdot 4\pi R/c \approx 4 \times 10^{-7}$ s. The two proper times are almost equal: $\tau_- \approx \tau_+ \approx 2\pi R/V \approx 40$ h. The

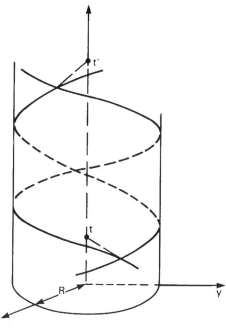

Fig. 3.3

relative difference is $(2V/c)(R\Omega/c) \approx 3 \times 10^{-12}$. Current clocks can achieve this accuracy and the effect has been measured.

3.2.3 The muon lifetime

Muons, once called "μ mesons", formed in the upper atmosphere, have velocities close to c. Their decay lifetime is $\tau_\mu \approx 2.2 \times 10^{-6}$ s. They are observed on the Earth, a possibility excluded by Newtonian theory but not by relativity.

Muons, pions with lifetimes $\tau_\pi \approx 2.6 \times 10^{-8}$ s, as well as other particles with very short lifetimes, are produced in particle accelerators and sometimes transported distances ℓ of several kilometres with speeds less than c.

The maximum distance a muon can travel is determined by the relation $\tau \approx \tau_\mu$ where τ is the proper time of the muon between formation and decay. As an exercise the reader should represent the motion of a muon in a space–time diagram and find its velocity for $\ell > 1$ km and $\ell > 30$ km.

> We can see how to measure the lifetime of an elementary particle from photographs like fig. 4.9, assuming that all the particles are identified and that the velocity of a charged particle can be measured by measuring the curvature of its trajectory. Consider the case of a charged particle created in a bubble chamber, in a collision or from the decay of another particle.
>
> Discuss the conditions required for the measurement when the lifetime is very short, or when the particle is neutral.

Fig. 3.4

The LEP and SPS experiments at CERN (cf. photograph on p. xiii), are subterranean. This photograph shows the underground corridor and part of the SPS ring, 50 metres below ground. Protons circulate in an evacuated tube, their tracks being curved by magnets placed at regular intervals along the tube. Between magnets the tracks are straight. Compare the geometry of the SPS (photographs on p. xiii and above) with that of the cyclotrons decribed in § 4.4.3.

(Photograph: CERN)

3.2.4 Length contraction, time dilation

We have avoided using expressions such as "length contraction" or "time dilation" in the relativistic context. Such expressions tend to distort the ideas of relativity. We can however *define* the length of a ruler in a reference system R or R'.

Consider a ruler stationary in the $\{x, t\}$ plane in R, whose ends A and B have worldlines $x_A = a$ and $x_B = b$, where a and b are constants.

We can thus characterise the world-tube T of the ruler in R and R', a Minkowski reference system related to R by a special Lorentz transformation.

Consider in R the intersection of the instant E_t defined in section 1.2 with T. Similarly, consider in R' the intersection of T and $E_{t'}$. These intersections define the "geometrical state" of the ruler in R at the instant t and in R' at the instant t'. The length of the ruler in R is then defined as $|x_A(t) - x_B(t)|$ and in an analogous way in R'. Using this definition, we find that the length of the ruler is less in R' than in R. In R' the ruler is moving, and is "shorter" than in R, where it is at rest.

To clarify the situation, we can represent in R the intersection of $E_{t'}$ and T, which is a spacelike interval. We can verify that its length in R' is the integral of the element of "proper length" $d\ell = \sqrt{-(c^2\, dt^2 - dx^2)}$ over this interval.

In a similar way, calling the difference of time coordinates "duration" in a given reference system, we can show that the duration between two events on the worldline of a clock is "dilated" when the clock moves.

"Contraction" of lengths and "dilation" of durations are not intrinsic physical phenomena, but connected with the definitions adopted.

Chapter 4

Dynamics of a particle

In both the Newtonian and relativistic theories, the motion of a point particle differs from that of a free particle when it is acted on by a **force**.

Two different questions arise:

(1) What is the nature and expression for the force acting on the particle?

(2) What is the relation between the motion of a particle and the force acting on it?

The answer to the first question depends on the physical conditions: which fields are present, the nature of the particle, the presence of friction, etc.

The second question has a very general answer; this is the subject of this chapter.

4.1 The equation of motion

4.1.1 Rest-frame

In studying the motion of a single particle it is always possible to consider a Minkowski reference system moving with the particle "for a short time" near an arbitrary event E^0. This reference system is called the **rest frame**.

In the rest frame at E^0, the particle's velocity remains small compared with c in a neighbourhood of E^0. We may then use the validity of Newtonian theory for slow motions to find the relativistic form of the dynamical laws.

Among the various rest frames, there is a particular one we shall call the "tangent frame". In this frame the four-velocity $\vec{U}$, the "velocity" $\overline{V}$, the "acceleration" $d\overline{V}/dt$ and the four-acceleration $\vec{\Gamma}$ have components

$$
\begin{cases}
U^\alpha = \{U^0, U^k\} = \{c, 2a\tau, 0, 0\} + O(\tau^2) \\
V^k = U^k + O(\tau^2), \quad dV^k/dt = \Gamma^k + O(\tau) \\
\dfrac{dU^\alpha}{d\tau} \equiv \Gamma^\alpha = \{\Gamma^0, \Gamma^k\} = \{0, 2a, 0, 0\} + O(\tau)
\end{cases}
\qquad (4.1.1)
$$

where k takes the values 1, 2, 3 and α the values 0, 1, 2, 3. In these expressions a is a positive constant and τ is the proper time measured along the wordline of the particle starting from the origin E^0. Further, we have set $O(\tau^n) = \tau^n \cdot f(\tau)$ where $f(\tau)$ is a bounded function of τ in the neighbourhood of $\tau = 0$.

Consider a particle whose motion in a Minkowski reference system is determined by the four functions $X^\alpha(\tau)$:

$$x^\alpha = X^\alpha(\tau).$$

On the particle's worldline, we consider the event E^0 corresponding to the value τ^0 of the parameter.

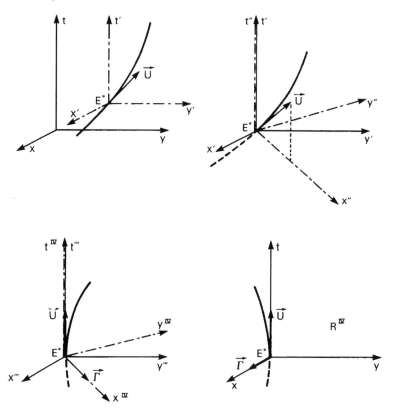

(1) We make a coordinate transformation:

$$x'^\alpha = x^\alpha - X^\alpha(\tau^0).$$

The event E^0 is then situated at the origin of the new coordinate system: $x'^\alpha(E^0) = 0$. In the new reference system the "velocity" at E^0 has components $V^{k'} = dx'^k/dt'$.

(2) We make the coordinate change corresponding to a global rotation of the axes such that in the new system we have at E^0,

$$V^{2''} = V^{3''} = 0.$$

(3) A special Lorentz tranformation reduces the velocity to zero at E^0 in the new reference system:

$$(dx'''^k/d\tau)_{E^0} = 0.$$

The latter reference system is a rest-frame. There exist several rest-frames at E^0.

(4) We can make another rotation of the axes so that in the new rest-frame R'''' we have

$$d^2 y''''/d\tau^2 = d^2 z''''/d\tau^2 = 0 \quad \text{and} \quad d^2 x''''/d\tau^2 > 0$$

at E_0.

Under these conditions, measuring the proper time from the origin E^0, $\tau^0 = 0$, we get

$$x = a \cdot \tau^2 + O(\tau^3), \quad t = \tau + O(\tau^3),$$
$$y = O(\tau^3), \quad z = O(\tau^3),$$

where a is a positive constant. Here we have used the notation x, y, z, t for the coordinates rather than $x'''', y'''', z'''', t''''$. We have written $O(\tau^n)$ for all quantities of the form $\tau^n \cdot f(\tau)$, where $f(\tau)$ is a bounded function of the variable τ near $\tau = 0$. Thus if τ remains fairly small, $O(\tau^m)$ is negligible compared with τ^n and $O(\tau^n)$ for $n < m$.

Differentiating $O(\tau^n)$ and assuming that $df/d\tau$ remains bounded, we get the relation $dO(\tau^n)/d\tau = O(\tau^{n-1})$.

We now use the definitions of U^α, Γ^α and V^k (cf. 1.4.1, 1.4.3 and 2.3.4), to get the expressions (4.1.1) above.

4.1.2 The equation of motion

In Newtonian mechanics, every particle is characterised by its mass m. Momentum is defined as the vector $\vec{p} = m \cdot \vec{V}$ where $\vec{V}$ is the velocity of the point mass with respect to the Galilean reference system considered. The fundamental relation of dynamics (the equation of motion) can then be expressed in the form $\vec{f}_N = d\vec{p}/dt = m \cdot \vec{\gamma}$ where $\vec{f}_N$ and $\vec{\gamma}$ are the Newtonian force acting on the point mass and its acceleration respectively.

To get the relativistic form of the fundamental relation we consider a Minkowski reference system R tangent at E^0 to the worldline of the particle considered. In R relativity reduces to Newtonian theory in the neighborhood of E^0. Using the relations (4.1.1) we get

$$f^k{}_N = \frac{dp^k}{dt} = m\frac{dV^k}{dt} = \frac{mdU^k}{d\tau} + O(\tau). \tag{4.1.2}$$

Let us define the four-vector $\vec{F}$ whose components in the tangent frame are

$$F^k = f^k{}_N \quad \text{and} \quad F^0 = 0.$$

This definition determines the components $F^{\alpha'}$ of $\vec{F}$ in any reference system starting from the Newtonian expressions in R.

We note that $dU^0/d\tau = 0$ at E^0; the fundamental relation (4.1.2) can then be written

$$F^\alpha = \frac{mdU^\alpha}{d\tau} = m \cdot \Gamma^\alpha. \tag{4.1.3}$$

These equations are the expression in a particular basis of an equality between four-vectors,

$$\vec{F} = m\vec{\Gamma}. \tag{4.1.4}$$

This vector equation is independent of the basis used to establish it, and holds for an arbitrary event E^0; it is therefore true generally.

More often, one may prefer to introduce the four-momentum, with components $P^\alpha = m \cdot U^\alpha$, and express the fundamental relation of dynamics in the form

$$\blacklozenge \qquad F^\alpha = \frac{dP^\alpha}{d\tau} \quad, \text{ with } P^\alpha = mU^\alpha. \qquad (4.1.5)$$

We note that the relations (1.4.2) and (1.4.4) imply

$$\blacklozenge \qquad (\vec{P}, \vec{P}) = m^2c^2 \quad \text{and} \quad (\vec{P}, \vec{F}) = 0 = (\vec{U}, \vec{F}). \qquad (4.1.6)$$

Thus the fundamental relation (4.1.5) of relativistic dynamics represents only three independent relations between the components of the four-momentum P^α and the four-force F^α.

4.1.3 Interpretation for slow motion

Using the expression (2.3.4) for the four-velocity and the expression for P^α we get

$$P^0 = \frac{mc}{\sqrt{1 - \beta^2}} \qquad P^k = \frac{mV^k}{\sqrt{1 - \beta^2}} \qquad (4.1.7)$$

$$V^k = \frac{cU^k}{U^0} = \frac{cP^k}{P^0}. \qquad (4.1.8)$$

We use the Newtonian notations introduced in section 2.3.2. Thus $\beta^2 = \overline{V}^2/c^2$ with $\overline{V}^2 = \sum_k V^k \cdot V^k$.

In this notation, using (4.1.6) and (4.1.8), we get

$$F^0 U^0 = \sum_k F^k \cdot U^k \quad \text{or} \quad cF^0 = \overline{F} \cdot \overline{V}. \qquad (4.1.9)$$

In the limit of low velocities, $\beta^2 \approx 0$, expanding (4.1.7) gives

$$cP^0 = mc^2 + \frac{1}{2}m\overline{V}^2, \quad \overline{P} = m\overline{V}.$$

We see that $c \cdot P^0$ is the kinetic energy E_k of the system, up to an additive constant: mc^2. This constant is called the **rest-mass energy** of the particle, or simply the "rest-energy".

The components of the Newtonian momentum are then P^k.

In the limit of small velocities, $F^k \approx f^k{}_N$ from (4.1.9); cF^0 is the power $\mathscr{P}$, delivered by the force $\vec{f}_N$ on the particle; $\vec{f}_N$ is a three-dimensional force of the Newtonian theory.

The four fundamental relations (4.1.5) are thus interpreted in the limit $\beta^2 \to 0$ as:

$$F^k = \frac{dP^k}{d\tau} \to \vec{f}_N = \frac{d\vec{p}}{dt} : \text{Newton's second law}$$

$$F^0 = \frac{dP^0}{d\tau} \to \mathscr{P} = \frac{dE_k}{dt} : \text{kinetic energy theorem.}$$

We know that the Newtonian kinetic energy theorem follows from Newton's second law, and does not express an independent relation. The situation is analogous in relativity, where only three equations are independent.

4.1.4 Zero rest-mass particles

We defined the four-momentum using the mass of the particle and its four-velocity. In reality, it is the four-momentum which is the fundamental quantity. The mass is defined from $\vec{P}$ by the relation (4.1.6).

It is convenient to assume that light is a flux of photons, particles moving at the speed of light c. The relation (4.1.8), with $\overline{V}^2 = c^2$ implies $\{\sum_k P^k \cdot P^k\}/(P^0)^2 = 1$, or $m = 0$: *particles of speed c have zero rest-mass.*

The converse holds: in any Minkowski reference system, a particle of zero rest-mass moves at speed c.

Thus, the relations $(\vec{P}, \vec{P}) = (P^0)^2 - \left\{\sum_k P^k \cdot P^k\right\} = m^2 = 0$ imply

$$\left\{\sum_k P^k \cdot P^k\right\}/(P^0)^2 = 1, \text{ or } \overline{V}^2 = c^2.$$

We note that for a zero rest-mass particle the relation $\beta^2 = 1$ implies $d\tau = 0$ (cf. 1.2.1). *This relation characterises zero rest-mass particles.*

The concept of a particle of zero rest-mass is a limiting concept, as such particles cannot be brought to rest. These particles are associated with waves, electromagnetic waves for photons for example. The energy E, defined by the relation $E = cP^0$, can be expressed as a function of the frequency v of the associated wave: $E = hv$, where h is Planck's constant. The vanishing mass then implies the relation $\|\overline{P}\| = hv/c$. The photoelectric effect, the expulsion of an electron from a metal by a photon, and the Compton effect, the collision of an electron and a photon, were the first experimental confirmations of these results.

4.2 Systems of particles

4.2.1 Pseudo-Newtonian formulation of the fundamental relation

To study the behaviour of several point masses in a given Minkowski reference system, it is useful to use the coordinate t as a parameter in preference to the τ parameters, which are specific to each of the particles.

A bubble chamber contains a liquid near its boiling point, liquid hydrogen at a temperature of order 20 K for example. As the beam of accelerated particles passes, the pressure is suddenly lowered. The charged particles passing through this medium, which has become metastable, leave a trail of ions behind them, which favour evaporation. Charged particles thus leave a trace in the form of a trail of small bubbles. Their tracks are curved by the magnetic field acting upon them. Several photographs are taken from different angles so that the trajectory in space can be reconstructed. Neutral particles leave no trace.

In the last few years bubble chambers have been abandoned in favour of other detectors, proportional counters for example.

A proportional counter consists of a set of intersecting wires in a gas-filled enclosure. Each wire is connected to a high-voltage source and an electronic circuit. A charged particle ionises the gas in the enclosure. Electrons are attracted by the closest wire, giving its position and allowing one to reconstruct its path.

At CERN, the UA1 detector is stationed at the point where beams of protons and antiprotons collide. It allows one to visualise the tracks of the various particles created in such collisions.

Fig. 4.1   Fig. 4.2

Fig. 4.1 shows the small BIBC bubble chamber. The scale is given by its diameter, which is about 10 cm. Much larger chambers have been constructed, such as HBC (used to take the pictures on pp. 47 and 50) which uses a hydrogen reservoir 2 m long, or Gargamelle, which was filled with a "heavy" liquid (propane–freon), and was involved in the discovery of neutral currents.

Fig. 4.2 shows the magnet of UA1 open, with the central detector removed. A comparison of these two photographs emphasises the tendency towards gigantic experiments in some areas of physics (see also the figure on p. 185 in a different area).

(Photographs: CERN)

Table 4.1

Definitions and useful expressions

$$V^k \equiv \frac{dx^k}{dt}, \quad \beta^2 \equiv \frac{\sum_k V^k \cdot V^k}{c^2}, \quad U^0 \equiv \frac{c}{\sqrt{1-\beta^2}}$$

$$f^\alpha \equiv \frac{cF^\alpha}{U^0}, \quad \text{where } F^\alpha \text{ is the four-force}$$

$$P^k \equiv \frac{mV^k}{\sqrt{1-\beta^2}}, \quad E \equiv cP^0 \equiv \frac{mc^2}{\sqrt{1-\beta^2}}$$

$$E \equiv \sqrt{m^2 c^4 + c^2 \overline{P}^2}$$

Fundamental dynamical relation

$$f^k = \frac{dP^k}{dt} \qquad\qquad \frac{dE}{dt} = \vec{f} \cdot \vec{V} \equiv \sum_k f^k \cdot V^k$$

With this aim, we define the "relativistic force" $f^\alpha = cF^\alpha/U^0$ at each point of the worldline of each particle.

In table 4.1 we summarise the definitions used as well as the form of the fundamental relation of dynamics when we use the "relativistic force".

Note that we have used the term "force" in three different contexts: $f^k{}_N$ denotes the Newtonian force, F^α denotes the four-force and f^α a relativistic force introduced for convenience. In a rest frame, the quantities $f^k{}_N$, f^k and F^k are equal.

As long as we do not change the Minkowski reference system, we can argue in Newtonian terms, provided we modify the fundamental dynamical relation, whose new form is given in table 4.1.

4.2.2 Systems of particles

We consider a system of free particles. In a given Minkowski reference system, we define the total four-momentum $\vec{P}$ at time t as the sum of the four-momenta of each free particle at that time:

$$\vec{P} = \sum_a \vec{P_a}, \tag{4.2.1}$$

where a is an index numbering the particles.

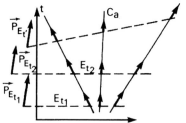
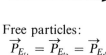
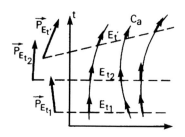

Free particles: Accelerated particles:
$$\vec{P}_{E_{t_1}} = \vec{P}_{E_{t_2}} = \vec{P}_{E_{t'}}$$ $$\vec{P}_{E_{t_1}} \neq \vec{P}_{E_{t_2}} \neq \vec{P}_{E_{t'}}$$

Fig. 4.3

The total momentum $\{P^k\}$ and total energy E of a set of particles with respect to the Minkowski reference system are then:

$$
\begin{cases}
P^k = \sum_a P_a^k, & \overline{P}_a = \dfrac{m_a \overline{V}_a}{\sqrt{1 - \left(\dfrac{V_a}{c}\right)^2}} \\[4ex]
cP^0 \equiv E = \sum E_a, & E_a = \dfrac{m_a c^2}{\sqrt{1 - \left(\dfrac{V_a}{c}\right)^2}}.
\end{cases}
\qquad (4.2.2)
$$

As the particles are free, P^k and E are constant.

To calculate $\vec{P}$ we select the event on each worldline $\mathscr{C}_a$ of each particle in E_t, the set of events at the same time t. The sum of the four-momenta associated with each of these events then gives $\vec{P}$. When we change the Minkowski reference system, we have to choose the events of $E_{t'}$, corresponding to the time t' in the new reference system. However, whatever the event considered on $\mathscr{C}_a$, the value of $\vec{P}_a$ is unchanged. The total four-momentum is thus independent of the reference system: when we change the system, the four-vector $\vec{P}$ is unchanged, but its components P^α change because we have changed the basis.

We can show that for a system of free particles, one can find a Minkowski reference system in which P^k vanishes. Such a reference system is called the **centre of mass frame**.

The situation is more complicated when forces act on the particles (fig. 4.3). In this case, $\vec{P}$ depends on the reference system R chosen and on the time t (we must use the notation $\vec{P}_{(R,t)}$ rather than $\vec{P}$; $\vec{P}_{(R,t)}$ is still a four-vector as it is a sum of four-vectors).

A particularly important case in physics is that of composite particles. These particles consist of point masses linked to each other by interactions

Leptons, or light particles (for example the electron and positron and the muon and antimuon) are fundamental constituents of matter, along with the quarks. Thus the proton and antiproton are composite particles made from quarks. High-energy collisions allow one to study the composition of these particles and to test theories of fundamental interactions (the strong interaction between hadrons, built out of quarks, or the electromagnetic interaction between charges, for example) and the associated field particles (the gluon and the photon).

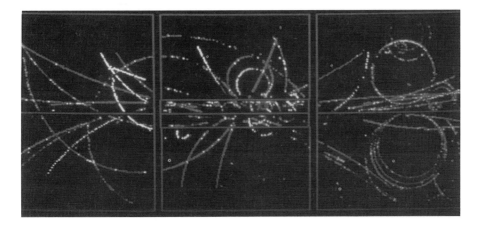

Fig. 4.4

The data collected by UA1 at CERN from proton–antiproton collisions allow computer reconstruction of the traces of the created particles. The photograph shows a longitudinal view of a reconstruction of this type.
(Photographs: CERN)

internal to the particle: for example the hydrogen atom formed from a proton and an electron linked by the Coulomb interaction. In some low-energy experiments, the atom behaves as a point with constant mass. Detailed study of such systems is complicated for various reasons, as the treatment of electromagnetic interactions is not simple, and more importantly, because we must use quantum mechanics. We do not discuss this subject here.

4.3 Collisions

We consider a system initially formed from a set of free particles. These particles approach each other closely enough to allow interactions of various types which can change the state of the system. These changes may be simple deflections, or they may change the nature of the particles present. In the final state the particles are far enough from each other that they constitute a new system of free particles.

Under these conditions, *the total four-momentum of the system is constant.* There are four laws governing collisions: the laws of *conservation of energy E and the momentum components P^k.*

These laws are extremely well verified experimentally. However, they do not completely determine the possible processes: they are necessary conditions only.

We shall limit this study to a few representative cases. The diagrams presented are "space" diagrams, and as an exercise we suggest that the reader represent the collisions studied on a space–time diagram $\{x, y, t\}$ by suppressing the z coordinate.

4.3.1 Elastic collisions

In such collisions, *the nature of the particles stays unchanged.*

Consider two particles of the same mass m, of which one is initially at rest and the other has initial velocity $\overline{V} = \{V, 0, 0\}$, $V > 0$ (figs. 4.5 and 4.6). After the collision, we assume (for example) that the two particles have velocities symmetric about the x-axis:

$$\overline{V}_1 = V_1\{\cos\theta, \sin\theta, 0\} \quad \text{and} \quad \overline{V}_2 = V_2\{\cos\theta, -\sin\theta, 0\}.$$

We set $\beta = V/c$, $\beta_1 = V_1/c$, $\beta_2 = V_2/c$.

Conservation of momentum for each axis gives the relations

for Ox :
$$\frac{mV}{\sqrt{1-\beta^2}} = \frac{mV_1 \cos\theta}{\sqrt{1-\beta_1^2}} + \frac{mV_2 \cos\theta}{\sqrt{1-\beta_2^2}}$$

for Oy :
$$0 = \frac{mV_1 \sin\theta}{\sqrt{1-\beta_1^2}} - \frac{mV_2 \sin\theta}{\sqrt{1-\beta_2^2}}$$

for Oz : $0 = 0$ identically satisfied.

Conservation of energy gives the relation

$$mc^2 + \frac{mc^2}{\sqrt{1-\beta^2}} = \frac{mc^2}{\sqrt{1-\beta_1^2}} + \frac{mc^2}{\sqrt{1-\beta_2^2}}.$$

The equations can be solved:
(1) In the "non-relativistic" case, $\beta \ll 1$,

$$V_1 = V_2 = V/\sqrt{2} \qquad\qquad \theta = 45°.$$

Quite generally, we can show in this case that the velocities of the particles after the collision are always orthogonal.
(2) In the relativistic case, $\beta = 0.9$ for example,

$$V_1 = V_2 \approx 0.8c \qquad\qquad \theta \approx 38°.$$

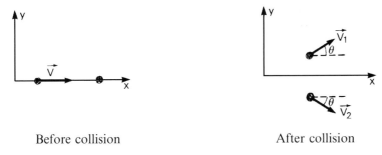

Before collision	After collision

Fig. 4.5

Before HBC was dismantled, almost 12 000 photographs were taken at CERN by this bubble chamber and analysed at LPNHE (Laboratoire de Physique Nucléaire et Hautes Energies – Universités de Paris VI and VII, and CNRS, Centre National de la Recherche Scientifique). The work was undertaken on the initiative and under the direction of J. Duboc, a member of this laboratory, so as to provide a selection of photographs based on precise illustrative criteria. The pictures below and on p. 50 are taken from this selection.

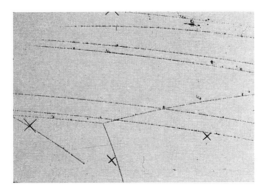

Fig. 4.6

Here a proton incident from the left ($v \approx 0.9c$) collides with a proton of one of the atoms of liquid hydrogen. This collision can be regarded as one between a projectile and a target of the same mass, which is initially at rest (cf. § 4.3.1). In contradiction to the Newtonian prediction, but in agreement with relativity, the particle velocities, tangent to their trajectories, are not orthogonal after the collision. Here this property can be seen directly as the plane of the trajectories is parallel to the plane of the film.

(Photograph: CERN)

Quite generally the angles observed in collisions between fast particles from accelerators are not those predicted by Newtonian theory but those given by relativity, which implies a test of the latter theory.

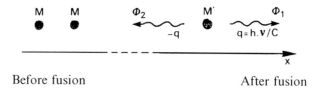

Before fusion   After fusion

Fig. 4.7

4.3.2 Fusion

Consider two particles of mass M_1 and M_2. The particles approach and fuse into a single particle of mass M'; some zero rest-mass particles may also be emitted, such as photons or neutrinos. Such phenomena occur for example when an ion absorbs an electron to form an atom, or when a nucleus absorbs a neutron.

To simplify the calculations we assume that the two masses are equal, $M_1 = M_2 = M$, and that they are initially almost at rest: $p \approx 0$. We further assume that two identical photons, Φ_1 and Φ_2, of energy cq are emitted in opposite directions along Ox (fig. 4.7); their momenta are thus q and $-q$ respectively.

As momentum is conserved, it is zero after the collision as it was zero before. As the photons have zero momentum, the mass M' is at rest. The conservation of total energy is then

$$Mc^2 + Mc^2 = 2cq + M'c^2.$$

We see immediately that the mass M' is less than the sum of the masses of its constituents. The difference is called the **mass defect**, and we write it as ΔM. The energy liberated in the fusion process is ΔMc^2. In the example under discussion this energy appears as radiation: photons. To split M' into its original constituents we would have to supply the energy ΔMc^2:

The mass defect is a measure of the binding energy of the system.

This energy is the ionisation energy of the outer electrons, of the order of a few electron volts per molecule, corresponding to a relative mass defect of less than 10^{-9}. We can see why the conservation of mass was for a long time regarded as a fundamental law. In nuclei, the mass defect does not exceed 10^{-2}, even for the most stable nuclei.

Analysis of the mass defects of the various nuclei shows that fission of "heavy" nuclei into "light" nuclei liberates energy: the internal energy of the heavy nuclei is greater than that of their fission products. This property is used in nuclear reactors. By contrast the fission of light nuclei requires energy, so that fusing light nuclei into heavier ones produces energy. These "thermo-nuclear" reactions occur in the stars, for example in the Sun. We are not as yet able to produce these reactions under controlled conditions to give useful energy. At the end of 1991 a controlled fusion reaction was produced for the first time by the European JET project. This result augurs well for the future.

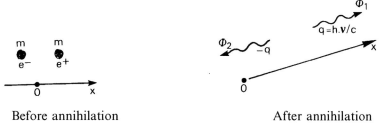

| Before annihilation | After annihilation |

Fig. 4.8

4.3.3 Annihilation of matter

We saw in the example above that the mass of a system depends on its internal energy. We consider here a collision in which all the mass disappears to produce radiation.

Consider an electron and a positron. These two particles form a particle–antiparticle pair. They have the same mass $m = 0.91 \times 10^{-30}$ kg and opposite charges.

We assume that the two particles intially have very small velocities along the axis Ox. As they approach each other under their electrostatic attraction, they annihilate to produce two photons Φ_1 and Φ_2.

As the momentum was initially zero, it remains zero. The two photons move along the Ox axis with opposite momenta. They therefore have the same energy $h\nu$ (fig. 4.8).

The conservation of energy gives the relation

$$2h\nu = 2mc^2.$$

We deduce the wavelength of the emitted radiation: $\lambda = h/mc \approx 0.024$ Å.

We note that the emission of at least two photons is required to satisfy the conservation laws.

We emphasise finally that in this experiment all of the rest-mass energy lost is transformed into radiation energy.

The mass, M, is a measure of the maximal internal energy E which can be recovered in any other form: $E = Mc^2$.

Conversely, all forms of energy are endowed with inertia: the mass of a composite system depends on the kinetic energy of its constituents and on the potential energy resulting from the interacting fields present. Indeed, we shall see that light is deflected by the gravitational field of the Sun, just like a mass.

4.4 Applications

As exercises for the reader we suggest the following.

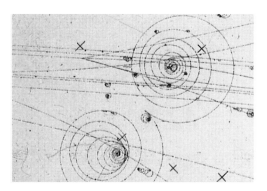

Fig. 4.9

At the top of the photograph a negatively-charged particle K^- coming from the left collides with the proton of an atom of liquid hydrogen in the bubble chamber. Various particles are created, including a neutral particle, π^0, which decays very quickly into two photons. One of these photons collides with a hydrogen atom, producing an electron–positron pair in the collision. These two particles spiral in opposite senses around the magnetic field. This reaction, transformation of radiation into matter, may be considered in a sense the inverse reaction to the annihilation of a particle–antiparticle pair into two photons (cf. § 4.3.3).

Considering figs. 4.6 and 4.9, it is useful to reflect on the role of the magnetic field, on the various quantities one can obtain from measurements of such photographs, as well as the comparison of Newtonian and relativistic theory they make possible.

(Photograph: CERN)

4.4.1 The Compton effect

Photons of wavelength λ and frequency ν collide with electrons of mass m, assumed at rest. The photons are scattered in all directions. Consider a particular collision in which the scattering angle between the initial and final direction of the photon is θ. Show that

$$\lambda' - \lambda = \frac{h(1 - \cos\theta)}{mc}$$

where λ' is the wavelength of the scattered photon and h Planck's constant. This effect showed that photons of energy $h\nu$ have momentum $h\nu/c$.

4.4.2 Energy production in fusion

Among other possible reactions, two deuterium nuclei can give a fusion reaction producing a tritium nucleus and a hydrogen nucleus and releasing 4.2 MeV. Deuterium is heavy hydrogen, present in the same compounds as hydrogen in atomic proportion of order 1.4×10^{-4}.

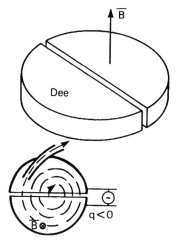

Fig. 4.10

A nuclear reactor produces a power of order 10^9 W. If one could harness the above reaction, show that it would produce the same power from the deuterium nuclei contained in a flow of about 0.3 l s^{-1} of water.

4.4.3 Particle acceleration

A cyclotron consists of a metal cylinder, in which a magnetic field $\overline{B}$, parallel to the axis of the cylinder, makes ions of charge q and mass m gyrate in circles in a plane perpendicular to $\overrightarrow{B}$ (fig. 4.10). The cylinder is in reality cut in two, so as to form two D-shaped semi-circular boxes called *dees*, separated by a slit, with a potential difference applied across them. This potential drop creates an electric field which accelerates the particle on each passage across the slit.

The relativistic force acting on the particle in an electric field $\overline{E}$ and a magnetic field $\overline{B}$ is

$$\overline{f} = q \cdot \overline{E} + q \cdot \overline{V} \wedge \overline{B}.$$

Assuming that $\overline{B}$ is constant in space and time, we can find the radius R of the trajectory as a function of the momentum P: $R = P/\|q\overline{B}\|$. The kinetic energy of the particle, $E_k = E - mc^2$, increases with each revolution.

For velocities small compared with c, the period T of the motion stays constant: $T = 2\pi m/\|q\overline{B}\|$. As the velocity increases the period changes and one has to synchronise the period of the applied potential difference. The cyclotron becomes a "synchrotron". We can express T and R as functions of the particle energy:

$$T = \frac{E}{mc^2} \cdot \frac{2\pi m}{\|q\overline{B}\|} \qquad R = \frac{\sqrt{E^2 - m^2 c^4}}{c\|q\overline{B}\|}.$$

The radius grows and the particle is ejected once it reaches the required energy.

4.4.4 Threshold energy

Particles can be produced in very high energy collisions. The antiproton, p^-, can be created in the collision of two protons: $p^+ + p^+ \longrightarrow 3p^+ + p^-$. We define Q as the difference between the rest-mass energy of the particles intially present and those of the particles produced. The mass of a particle is equal to that of its antiparticle; thus one can show that in the reaction considered $Q \approx -1.9$ GeV $= -1.9 \times 10^9$ eV.

One of the protons is at rest in the laboratory reference system R_L, assumed Minkowskian. We consider the centre of mass frame R_I in which the total momentum vanishes. To produce the reaction considered, we have to give the protons an energy E_L or E_I depending on what frame we work in. In R_L, the total energy is $E = 2mc^2 + E_L$ and the total momentum is P, with $(mc + E_L/c)^2 - P^2 = m^2c^2$. In R_I, the total energy is $E' = 2mc^2 + E_I$, while the total momentum is zero. The total four-momentum is a four-vector whose pseudo-length is independent of the reference system used to measure it, so we have $(E/c)^2 - P^2 = (E'/c)^2$. We can thus show that $2mc^2 E_L = E_I{}^2 + 4mc^2 E_I$. The minimum energy which must be supplied is called the **threshold energy**. Using conservation of energy and momentum in the collision, we can show that the threshold energy is $-Q = 2mc^2$ in R_I. The relation $E_I \geqslant -Q$ fixes the threshold energy in R_L: $E_L \geqslant 6mc^2$. It is easy to see the point of working in R_I. This is the point of colliding particle rings, such as the SPS at CERN (cf. the photographs on pp. xiii and 35) where protons and antiprotons are accelerated so that the laboratory reference system is precisely the centre of mass frame at the moment of collision.

Chapter 5

Mathematical framework

This and the following chapters of the first part of the book are not required in order to follow the second part.

The description we have introduced implies mathematical structure which we wish to make explicit.

5.1 Calculation rules for indices

We have used two kinds of indices, Latin ones such as i, j, k, etc., taking values 1, 2, 3, and Greek indices α, β, etc., running from 0 to 3.

5.1.1 The summation convention for repeated indices

Suppose that we have defined a quantity $T^{\alpha\beta j}{}_{\tau k}$. We wish to calculate the sum

$$\sum_{\alpha} T^{\alpha\beta j}{}_{\alpha k} \equiv T^{0\beta j}{}_{0k} + T^{1\beta j}{}_{1k} + T^{2\beta j}{}_{2k} + T^{3\beta j}{}_{3k}.$$

The index α is repeated twice on the left-hand side, and is called a **dummy index**; we could equally well replace it by τ as it does not take a single specified value, but all the values in the summation. The **Einstein summation convention** consists of dropping the $\sum$ symbol and writing

$$T^{\alpha\beta j}{}_{\alpha k} \equiv T^{0\beta j}{}_{0k} + T^{1\beta j}{}_{1k} + T^{2\beta j}{}_{2k} + T^{3\beta j}{}_{3k}.$$

Whether an index is Greek or Latin fixes the set of values it takes in the summation; for example

$$T^{\alpha\beta k}{}_{\tau k} \equiv T^{\alpha\beta 1}{}_{\tau 1} + T^{\alpha\beta 2}{}_{\tau 2} + T^{\alpha\beta 3}{}_{\tau 3}.$$

This rule applies only when the dummy index is repeated twice, once up and once down. We then say that we have peformed a **contraction** of T over the dummy index.

In performing a double summation we must use two different letters for the two dummy indices, for example:

$$
\begin{aligned}
A^{\alpha\beta}{}_{\alpha\beta k} \equiv A^{0\beta}{}_{0\beta k} \quad &\equiv A^{00}{}_{00k} + A^{01}{}_{01k} + A^{02}{}_{02k} + A^{03}{}_{03k} \\
+ A^{1\beta}{}_{1\beta k} \quad &+ A^{10}{}_{10k} + A^{11}{}_{11k} + A^{12}{}_{12k} + A^{13}{}_{13k} \\
+ A^{2\beta}{}_{2\beta k} \quad &+ A^{20}{}_{20k} + A^{21}{}_{21k} + A^{22}{}_{22k} + A^{23}{}_{23k} \\
+ A^{3\beta}{}_{3\beta k} \quad &+ A^{30}{}_{30k} + A^{31}{}_{31k} + A^{32}{}_{32k} + A^{33}{}_{33k}.
\end{aligned}
$$

The quantity obtained after the summation is characterised by the remaining indices other than dummy ones. Thus $T^{\alpha\beta j}{}_{\tau k}$ is a quantity with $4\cdot4\cdot3\cdot4\cdot3 = 576$ components while $T^{\alpha\beta k}{}_{\tau k}$ is a quantity with only $4\cdot4\cdot4 = 64$ components. We can for example write $T^{\alpha\beta k}{}_{\tau k} = A^{\alpha\beta}{}_{\tau}$. Finding the same non-dummy indices in the same place on each side of an equation is called **balancing the indices**

We can easily verify the following rules:

$$
\mathbf{A}^{\alpha}(\mathbf{N}_{\alpha} + q\mathbf{M}_{\alpha}) = \mathbf{A}^{\alpha}\mathbf{N}_{\alpha} + q\mathbf{A}^{\alpha}\mathbf{M}_{\alpha}
$$
$$
\mathbf{A}_{\alpha}(\mathbf{N}^{\alpha} + q\mathbf{M}^{\alpha}) = \mathbf{A}_{\alpha}\mathbf{N}^{\alpha} + q\mathbf{A}_{\alpha}\mathbf{M}^{\alpha}
$$

where $\mathbf{A}^{\alpha}$, $\mathbf{N}_{\alpha}$, etc. represent quantities possessing an arbitrary number of indices besides α, and q is an arbitrary number.

We can verify that the order of contraction does not matter:

$$
(\mathbf{A}^{\alpha}\mathbf{M}_{\alpha}{}^{\beta})\mathbf{N}_{\beta} = \mathbf{A}^{\alpha}(\mathbf{M}_{\alpha}{}^{\beta}\mathbf{N}_{\beta}).
$$

The summation convention also applies to operators.
Quite generally we can use the following notations:

$$
\begin{cases}
\dfrac{\partial(F)}{\partial x^{\alpha}} \equiv \partial_{\alpha}(F) \equiv (F)_{,\alpha} \\
x^{0} \equiv ct \quad x^{1} \equiv x \quad x^{2} \equiv y \quad x^{3} \equiv z.
\end{cases}
$$

Thus the divergence operator applied to a vector field in three-dimensional space, with components $V^{k}(x, y, z, t)$, is written

$$
\operatorname{div}[\vec{V}] \equiv \partial_{k}V^{k} \equiv (V^{k})_{,k}.
$$

The coordinates x^{α} will take a more general meaning below, as we shall abandon Minkowski reference systems. However, *the notations defined here will be maintained.*

The Einstein summation convention considerably simplifies writing.

5.1.2 Matrix; inverse matrix

Consider an array of numbers with two indices of the same type, Greek or Latin. Such an array defines a square matrix. Let us consider the case of Greek indices; four forms are possible: $A^{\alpha\beta}$, $A^{\alpha}{}_{\beta}$, $A_{\alpha\beta}$, $A_{\alpha}{}^{\beta}$.

To represent these matrices in the usual form of an array, we choose the following conventions: *the first index is that of the row and the second that of the column.*

The product $[P]$ of two matrices $[A]$ and $[B]$ is for example written

$$P^{\alpha}{}_{\sigma} = ([A] \cdot [B])^{\alpha}{}_{\sigma} = A^{\alpha\beta} B_{\beta\sigma}.$$

We note that in this form the Einstein summation convention applies in calculating $[A] \cdot [B]$ and $[B] \cdot [A]$. By contrast, if we consider the matrix $B_{\beta}{}^{\sigma}$, we see that $[B] \cdot [A]$ cannot be expressed using the Einstein convention, as the indices are both "up". As an exercise one can determine, given the form of $[A]$, all the possible forms of $[B]$ allowing calculation of $[A] \cdot [B]$ and $[B] \cdot [A]$ using the Einstein convention.

Whatever the form of $[A]$, we can calculate its determinant: we write it as Det$[A]$. If Det$[A]$ is non-zero we can find the inverse matrix to $[A]$. As an example, consider $[A]$ in the form $A^{\alpha}{}_{\beta}$. We give its inverse $[B]$ the form allowing us to use the Einstein convention in calculating $[A] \cdot [B]$ and $[B] \cdot [A]$. The form of $[B]$ is thus $B^{\alpha}{}_{\beta}$.

We define the **Kronecker delta**, $\delta^{\alpha}{}_{\beta} = \delta_{\beta}{}^{\alpha}$, so that

$$\delta^{\alpha}{}_{\beta} = \delta_{\beta}{}^{\alpha} = \{1 \text{ for } \alpha = \beta \text{ and } 0 \text{ for } \alpha \neq \beta\}.$$

The matrix $[B]$, inverse to the matrix $[A]$, thus satisfies the defining relations

$$A^{\alpha}{}_{\beta} B^{\beta}{}_{\sigma} = \delta^{\alpha}{}_{\sigma} \text{ or } \delta^{\alpha}{}_{\beta} = B^{\alpha}{}_{\beta} A^{\beta}{}_{\sigma}.$$

We recall that only one of these relations is needed to define $[B]$, the other relation being then automatically satisfied.

As an exercise the reader can write down the relations defining the matrices inverse to $A^{\alpha\beta}$, $A_{\alpha\beta}$, $A_{\alpha}{}^{\beta}$.

Given a matrix $[A]$ of the form $A^{\alpha}{}_{\beta}$ or $A_{\alpha}{}^{\beta}$, we define its **trace** as Tr$[A] = A^{\alpha}{}_{\alpha}$ or Tr$[A] = A_{\alpha}{}^{\alpha}$ in each case.

5.2 The dual space

5.2.1 Linear forms

Consider a vector space E, of dimension n, constructed over the field K of real or complex numbers.

We denote as a linear form on E a function $\mathbf{A}$ defined over E taking values in the field K, $\mathbf{A}[\vec{V}] \in K$, satisfying the relation

$$\mathbf{A}[\vec{V} + \mu \cdot \vec{W}] = \mathbf{A}[\vec{V}] + \mu \cdot \mathbf{A}[\vec{W}] \tag{5.2.1}$$

where $\vec{V}$ and $\vec{W}$ are arbitrary vectors of E and μ any element of K.

The form $\mathbf{A}$ is completely known if we know its effect on all the vectors of an arbitrary basis $\{\vec{e}_{\alpha}\}$ of E. We define $A_{\alpha} = \mathbf{A}[\vec{e}_{\alpha}]$. Given the vector $\vec{V}$ in the form $\vec{V} = V^{\beta} \cdot \vec{e}_{\beta}$, we have

$$\mathbf{A}[\vec{V}] = A_{\alpha} V^{\alpha} \quad \text{with} \quad A_{\alpha} \equiv \mathbf{A}[\vec{e}_{\alpha}] \tag{5.2.2}$$

5.2.2 The dual space E^* of E

We define the sum of two linear forms, **A** and **B**, and multiplication by an element μ of K:

$$(\mathbf{A} + \mathbf{B})[\vec{V}] = \mathbf{A}[\vec{V}] + \mathbf{B}[\vec{V}]$$
$$(\mu \cdot \mathbf{A})[\vec{V}] = \mu \cdot \mathbf{A}[\vec{V}]$$

for each vector $\vec{V}$ of E.

With these rules, the set of linear forms constitutes a vector space over K, **the dual space** of E, written E^*.

Given a basis $\{\vec{e}_\alpha\}$ of E, we define the set of n linear forms $\mathbf{e}^\alpha$ determined by their effect in the vectors $\vec{e}_\beta$ in the following way:

$$\mathbf{e}^\alpha[\vec{e}_\beta] = \delta^\alpha{}_\beta \tag{5.2.3}$$

One can easily verify the relation $\mathbf{A} = A_\alpha \cdot \mathbf{e}^\alpha$. Thus the $\{\mathbf{e}^\alpha\}$ form a basis of E^*: this is the **dual basis** to $\{\vec{e}_\alpha\}$; E^* and E have the same dimension.

We note that the dual basis is written with the index raised.

In the following, we shall consider only the case where E is the vector space T_4 introduced in section 1.3; K is then the field of real numbers, $\mathbb{R}$.

5.2.3 The scalar product in T_4

In section 1.3 we introduced a scalar product in T_4. As the basis of T_4 used here is not necessarily Minkowskian, we set

$$(\vec{e}_\alpha, \vec{e}_\beta) = (\vec{e}_\beta, \vec{e}_\alpha) = g_{\alpha\beta} = g_{\beta\alpha}.$$

Let $\vec{A}$ be a vector of T_4, chosen once and for all. The scalar product $(\vec{A}, \vec{V})$ defines a linear form on T_4 with values in the set of reals. This linear form associated with the vector $\vec{A}$ is written **A**, using the same letter:

$$\mathbf{A}[\vec{V}] \equiv (\vec{A}, \vec{V}) \in \mathbb{R}. \tag{5.2.4}$$

Given the decomposition of $\vec{A}$ over the basis $\{\vec{e}_\alpha\}$, we easily find the decomposition of **A** over the dual basis $\{\mathbf{e}^\alpha\}$:

$$\blacklozenge \qquad \begin{cases} \vec{A} = A^\alpha \cdot \vec{e}_\alpha \qquad \mathbf{A} \equiv A_\alpha \cdot \mathbf{e}^\alpha \\ A_\alpha = g_{\alpha\beta} A^\beta. \end{cases} \tag{5.2.5}$$

Thus $A_\alpha = \mathbf{A}[\vec{e}_\alpha] = (\vec{A}, \vec{e}_\alpha) = (A^\beta \cdot \vec{e}_\beta, \vec{e}_\alpha) = A^\beta \cdot g_{\beta\alpha} = g_{\alpha\beta} \cdot A^\beta$.

We introduce the inverse, $g^{\alpha\beta}$ of $g_{\alpha\beta}$:

$$\blacklozenge \qquad g^{\alpha\beta} g_{\beta\sigma} = \delta^\alpha{}_\sigma = \delta_\sigma{}^\alpha = g_{\sigma\beta} g^{\beta\alpha}. \tag{5.2.6}$$

We note that $g_{\alpha\beta} = g_{\beta\alpha}$ represents a symmetric matrix, as does $g^{\alpha\beta}$: $g^{\alpha\beta} = g^{\beta\alpha}$.

Contracting the two sides of (5.2.5) with $g^{\sigma\alpha}$, we obtain the components of $\overrightarrow{A}$ as a function of the components of **A** in the dual basis:

$$A^{\sigma} = g^{\sigma\alpha} A_{\alpha}. \qquad (5.2.7)$$

Vectors in T_4 are called **contravariant vectors** and those in T_4^*, **covariant vectors**.

The expressions (5.2.5) and (5.2.7) define a *one–one relation between T_4 and T_4^**. We can thus refer without ambiguity to the "covariant component" and "contravariant component" of the vector $\overrightarrow{A}$ to denote A_{α} and A^{α}.

As an exercise show that the covariant vector associated with $\vec{e}_{\alpha}$ by the relation above is the vector $\mathbf{e}_{\alpha} = g_{\alpha\beta} \cdot \mathbf{e}^{\beta}$.

5.3 Change of basis

5.3.1 Vectors of T_4

We consider the vector space T_4 of special relativity, with the basis $\{\vec{e}_{\alpha}\}$.
Let us make a change of basis:

$$\vec{e}_{\alpha'} = a^{\beta}{}_{\alpha'} \cdot \vec{e}_{\beta} \qquad (5.3.1)$$

with $\mathrm{Det}[a] \neq 0$.

We note that the "prime" characterising the new basis is applied to the indices: α becomes α'. This notation will prove very convenient.

Any vector $\overrightarrow{V}$ of E may be decomposed in either of the two bases introduced:

$$V^{\alpha} \cdot \vec{e}_{\alpha} = \overrightarrow{V} = V^{\alpha'} \cdot \vec{e}_{\alpha'} = V^{\alpha'} a^{\beta}{}_{\alpha'} \cdot \vec{e}_{\beta}; \qquad (5.3.2)$$

the notation of dummy indices has no significance, so we write

$$V^{\alpha'} a^{\beta}{}_{\alpha'} \cdot \vec{e}_{\beta} = V^{\alpha} \cdot \vec{e}_{\alpha} = V^{\beta} \cdot \vec{e}_{\beta}.$$

Identifying the components as the coefficients of $\vec{e}_{\beta}$ on each side we get

$$V^{\beta} = a^{\beta}{}_{\alpha'} V^{\alpha'}. \qquad (5.3.3)$$

Introduce the inverse matrix to $a^{\beta}{}_{\alpha'}$, which we write $a^{\alpha'}{}_{\beta}$:

$$a^{\alpha'}{}_{\beta} a^{\beta}{}_{\sigma'} = \delta^{\alpha'}{}_{\sigma'} \qquad a^{\alpha}{}_{\beta'} a^{\beta'}{}_{\sigma} = \delta^{\alpha}{}_{\sigma}. \qquad (5.3.4)$$

Contracting each side of (5.3.3) with $a^{\sigma'}{}_{\beta}$, we get

$$a^{\sigma'}{}_{\beta} V^{\beta} = a^{\sigma'}{}_{\beta} a^{\beta}{}_{\alpha'} V^{\alpha'} = \delta^{\sigma'}{}_{\alpha'} V^{\alpha'}$$

or

$$V^{\sigma'} = a^{\sigma'}{}_{\beta} V^{\beta}. \qquad (5.3.5)$$

The expressions (5.3.3) and (5.3.5) are the formulae for a change of basis in T_4.

5.3.2 Vectors of T_4^*

Consider the dual, T_4^*, of T_4. We set

$$a^{\alpha'}{}_{\beta} \cdot \mathbf{e}^{\beta} = \mathbf{e}^{\alpha'}. \qquad (5.3.6)$$

We may then easily show that $\{\mathbf{e}^{\alpha}\}$ is the basis dual to $\{\vec{e}_{\alpha'}\}$.

We express $\mathbf{e}^{\alpha'}[\vec{e}_{\beta'}]$ as a function of $\mathbf{e}^{\alpha}$ and $\vec{e}_{\beta}$ using $a^{\alpha'}{}_{\beta}$ and $a^{\alpha}{}_{\beta'}$. Using the defining properties of linear forms, we find $\mathbf{e}^{\alpha'}[\vec{e}_{\beta'}] = \delta^{\alpha'}{}_{\beta'}$.

Every covariant vector of T_4^* can be decomposed in the two bases, $\{\mathbf{e}^{\alpha}\}$ and $\{\mathbf{e}^{\alpha'}\}$: $\mathbf{A} = A_{\alpha} \cdot \mathbf{e}^{\alpha} = A_{\alpha'} \cdot \mathbf{e}^{\alpha'}$.
These relations lead to the equations

$$A_{\alpha'} = A_{\beta} a^{\beta}{}_{\alpha'} \quad \text{and} \quad A_{\alpha} = A_{\beta'} \cdot a^{\beta'}{}_{\alpha}. \qquad (5.3.7)$$

We adopt the convention of always using dual bases to represent T_4 and T_4^*. A change of basis is thus characterised by the expressions (5.3.3), (5.3.5) and (5.3.7).

5.3.3 The scalar product

The space T_4 has a scalar product:

$$g_{\alpha\beta} = (\vec{e}_{\alpha}, \vec{e}_{\beta}) \qquad g_{\alpha'\beta'} = (\vec{e}_{\alpha'}, \vec{e}_{\beta'}).$$

Replacing $\vec{e}_{\sigma'}$ by its expression as a function of $\vec{e}_{\sigma}$, we get

$$\begin{cases} g_{\alpha'\beta'} = g_{\sigma\mu} a^{\sigma}{}_{\alpha'} a^{\mu}{}_{\beta'} \\ g_{\alpha\beta} = g_{\sigma'\mu'} a^{\sigma'}{}_{\alpha} a^{\mu'}{}_{\beta}. \end{cases} \qquad (5.3.8)$$

The transformation formulae for the inverse matrix $g_{\alpha\beta}$ are given in the form

$$\begin{cases} g^{\alpha'\beta'} = a^{\alpha'}{}_{\sigma} a^{\beta'}{}_{\mu} g^{\sigma\mu} \\ g^{\alpha\beta} = a^{\alpha}{}_{\sigma'} a^{\beta}{}_{\mu'} g^{\sigma'\mu'}. \end{cases} \qquad (5.3.9)$$

We can verify explicitly that $g^{\alpha'\beta'}$ satisfies the conditions required of the inverse of $g_{\alpha'\beta'}$.

As an exercise, we can also verify the relation

$$A_{\alpha'} = g_{\alpha'\beta'}A^{\beta'}.$$

5.3.4 Changes of coordinates

Consider an arbitrary reference system for M_4, $\{O, \vec{e}_\alpha\}$, not necessarily Minkowskian. Let O' be a point of M_4 with coordinates b^α and $\{\vec{e}_{\alpha'}\}$ an arbitrary basis of T_4 such that

$$\vec{e}_{\alpha'} = a^\beta_{\alpha'} \cdot \vec{e}_\beta \qquad \vec{e}_\beta = a^{\alpha'}_\beta \cdot \vec{e}_{\alpha'} \tag{5.3.10}$$

when $a^\beta_{\alpha'}$ and $a^{\alpha'}_\beta$ are two matrices inverse to each other.

We consider the new reference system $\{O', \vec{e}_{\alpha'}\}$. Let A be an arbitrary event in M_4 with coordinates x^α in the first reference system. The natural coordinates associated with the new system are the components of $\overrightarrow{O'A}$ in the new basis: $\overrightarrow{OA} = x^\alpha \cdot \vec{e}_\alpha$ and $\overrightarrow{O'A} = x^{\alpha'} \cdot \vec{e}_{\alpha'}$. The relations $\overrightarrow{O'A} = \overrightarrow{O'O} + \overrightarrow{OA} = \overrightarrow{OA} - \overrightarrow{OO'}$ and $\overrightarrow{OO'} = b^\alpha \cdot \vec{e}_\alpha$ give the expression for the change of coordinates:

$$\begin{cases} x^\alpha = a^\alpha_{\beta'} x^{\beta'} + b^\alpha \\ x^{\alpha'} = a^{\alpha'}_\beta x^\beta - b^{\alpha'} \end{cases} \tag{5.3.11}$$

with $b^{\alpha'} = a^{\alpha'}_\beta b^\beta$.

Conversely, to every change of coordinates defined by either of the two expressions (5.3.11), there corresponds a change of the associated natural basis (5.3.10).

We can easily verify the expression below, where $d\tau$ is the proper time between two neighbouring events with coordinates $\{x^\alpha\}$ and $\{x^\alpha + dx^\alpha\}$ or $\{x^{\alpha'}\}$ and $\{x^{\alpha'} + dx^{\alpha'}\}$ depending on the coordinate system used:

$$c^2 d\tau^2 = g_{\alpha\beta}dx^\alpha dx^\beta = g_{\alpha'\beta'}dx^{\alpha'} dx^{\beta'}. \tag{5.3.12}$$

5.4 Tensor product

5.4.1 Definitions

Consider two vector spaces E and F with dimensions n and m respectively, constructed over the same field, K. In physical applications K is the real or complex numbers. Consider a third vector space H of dimension p, constructed over K.

H is called the **tensor product** of the spaces E and F and is written $E \otimes F$, if and only if with every pair of vectors $\{\vec{U}, \vec{V}\}$, $\vec{U} \in E$ and $\vec{V} \in F$, we can associate a vector $\vec{W}$ of H written $\vec{U} \otimes \vec{V} = \vec{W}$ such that

(1) if $\{\vec{e}_\alpha\}$ and $\{\vec{f}_k\}$ are two bases E and F, $\{\mathbf{h}_{\alpha k} = \vec{e}_\alpha \otimes \vec{f}_k\}$ is a basis of H (where $\alpha = 1, 2, \ldots, n$ and $k = 1, 2, \ldots, m$);

(2) if μ is any element of K, and $\vec{U}$, $\vec{U}_1$ and $\vec{U}_2$ are arbitrary vectors of E and $\vec{V}$, $\vec{V}_1$ and $\vec{V}_2$ arbitrary vectors of F, we have

$$\vec{U} \otimes (\vec{V}_1 + \vec{V}_2) = \vec{U} \otimes \vec{V}_1 + \vec{U} \otimes \vec{V}_2$$

$$(\vec{U}_1 + \vec{U}_2) \otimes \vec{V} = \vec{U}_1 \otimes \vec{V} + \vec{U}_2 \otimes \vec{V}$$

$$(\mu \cdot \vec{U}) \otimes \vec{V} = \vec{U} \otimes (\mu \cdot \vec{V}) = \mu \cdot (\vec{U} \otimes \vec{V}).$$

The operation $\otimes$ is called the **tensor product of two vectors**.

The first property implies that the dimension of $E \otimes F$ is $p = n \cdot m$. The second property allows us to calculate the tensor product of two vectors whose components with respect to the bases $\{\vec{e}_\alpha\}$ and $\{\vec{f}_k\}$ are known.

Consider two other bases $\{\vec{e}_{\alpha'}\}$ and $\{\vec{f}_{k'}\}$ of E and F. Set $\mathbf{h}_{\alpha'k'} = \vec{e}_{\alpha'} \otimes \vec{f}_{k'}$. Writing the components of the initial basis vectors with respect to (wrt) the new bases, we deduce that $\mathbf{h}_{\alpha k}$ is a linear combination of $\mathbf{h}_{\alpha'k'}$. $\{\mathbf{h}_{\alpha'k'}\}$ is a basis for H; thus if the first property holds for particular bases, it holds for arbitrary bases of E and F.

We call any vector of $E \otimes F$ a **tensor**. The components of any tensor with respect to the basis $\mathbf{h}_{\alpha k}$ are

$$\mathbf{T} = T^{\alpha k} \cdot \mathbf{h}_{\alpha k} = T^{\alpha k} \cdot \vec{e}_\alpha \otimes \vec{f}_k.$$

Defining $H = E \otimes F$, we note that we may define $H \otimes G = E \otimes F \otimes G$ where G is a vector space. We can use the procedure above quite generally to define the tensor product of an arbitrary number of vector spaces.

5.4.2 Changes of basis

Consider new bases $\{\vec{e}_{\alpha'}\}$ and $\{\vec{f}_{k'}\}$ of E and F:

$$\vec{e}_{\alpha'} = a^\beta{}_{\alpha'} \cdot \vec{e}_\beta \qquad \vec{f}_{k'} = b^j{}_{k'} \cdot \vec{f}_j$$

with $\text{Det}[a] \neq 0$ and $\text{Det}[b] \neq 0$.

Consider a tensor $\mathbf{T}$ of $E \otimes F$. Following the method used for vectors, we write

$$\mathbf{T} = T^{\beta j} \cdot \vec{e}_\beta \otimes \vec{f}_j = T^{\alpha'k'} \cdot \vec{e}_{\alpha'} \otimes \vec{f}_{k'} = T^{\alpha'k'} \cdot (a^\beta{}_{\alpha'} \cdot \vec{e}_\beta) \otimes (b^j{}_{k'} \cdot \vec{f}_j).$$

These relations give the formulae for transformation of the components of $\mathbf{T}$ under a change of basis:

$$\begin{cases} T^{\beta j} = a^\beta{}_{\alpha'} b^j{}_{k'} T^{\alpha'k'} \\ T^{\alpha'k'} = a^{\alpha'}{}_\beta b^{k'}{}_j T^{\beta j}. \end{cases} \tag{5.4.1}$$

Consider the case where F and E are identical. We define **symmetric tensors** as those whose components satisfy the relations $T^{\alpha\beta} = T^{\beta\alpha}$. Similarly tensors satisfying the relations $T^{\alpha\beta} = -T^{\beta\alpha}$ are called **antisymmetric**.

Using these formulae for changes of basis we can verify that these properties are independent of the basis; they therefore characterise the tensor itself.

5.4.3 Application to $\otimes^n T_4 \otimes^m T_4^*$

Here we consider only tensor products of T_4 and T_4^* in any number and order. The formalism above can be applied unaltered to tensor products of T_4. For tensor products of T_4^* the indices of the basis vectors are *raised* while those of the components are *lowered*. We make the notation precise by considering the tensor product $T_4 \otimes T_4^*$.

Let $\{\vec{e}_\alpha\}$ be a basis of T_4 and $\{\mathbf{e}^\alpha\}$ the dual basis of T_4^*. Let $\mathbf{T}$ be a tensor of $T_4 \otimes T_4^*$.

$$\mathbf{T} = T^\alpha{}_\beta \cdot \vec{e}_\alpha \otimes \mathbf{e}^\beta.$$

Now make the change of basis (5.3.1) and (5.3.6), which takes us from the frame R to the frame R':

$$\mathbf{T} = T^{\alpha'}{}_{\beta'} \cdot \vec{e}_{\alpha'} \otimes \mathbf{e}^{\beta'}.$$

In the first expression for $\mathbf{T}$, we can write the vectors $\vec{e}_\alpha$ and $\mathbf{e}^\beta$ as functions of the new basis:

$$\mathbf{T} = T^\alpha{}_\beta \cdot (a^{\sigma'}{}_\alpha \cdot \vec{e}_{\sigma'}) \otimes (a^\beta{}_{\mu'} \cdot \mathbf{e}^{\mu'}) = T^{\alpha'}{}_{\beta'} \cdot \vec{e}_{\alpha'} \otimes \mathbf{e}^{\beta'}.$$

Identifying the coefficients of the basis vectors $\vec{e}_{\rho'} \otimes \mathbf{e}^{\tau'}$ on each side we get

$$T^{\alpha'}{}_{\beta'} = a^{\alpha'}{}_\tau a^\mu{}_{\beta'} T^\tau{}_\mu.$$

This expression generalises immediately to arbitrary tensors with the indices in arbitrary positions; for example for a tensor of $T_4 \otimes T_4^* \otimes T_4$: $\mathbf{T} = T^\alpha{}_\beta{}^\tau \cdot \vec{e}_\alpha \otimes \mathbf{e}^\beta \otimes \vec{e}_\tau$. The rule is as follows:

(1) in front of $T^\alpha{}_\beta{}^\tau$ we write $a^-{}_-$ as many times as it has indices,

(2) on each of the $a^-{}_-$ we write an index α, β, τ, either raised or lowered, so as to allow contraction,

(3) we complete the expression by placing on each of the $a^-{}_-$ the primed indices $\mu', \sigma', \varepsilon'$. The rule of balanced indices indicates the component thus expressed.

The sequence of operations for expressing the components of $\mathbf{T}$ in the new basis is thus:

$$T^\alpha{}_\beta{}^\tau \longrightarrow (1)\ a^-{}_- \cdot a^-{}_- \cdot a^-{}_- \cdot T^\alpha{}_\beta{}^\tau \longrightarrow (2)\ a^-{}_\alpha \cdot a^\beta{}_- \cdot a^-{}_\tau \cdot T^\alpha{}_\beta{}^\tau$$

$$\longrightarrow (3)\ a^{\sigma'}{}_\alpha \cdot a^\beta{}_{\mu'} \cdot a^{\varepsilon'}{}_\tau \cdot T^\alpha{}_\beta{}^\tau = T^{\sigma'}{}_{\mu'}{}^{\varepsilon'}$$

$$T^{\sigma'}{}_{\mu'}{}^{\varepsilon'} = a^{\sigma'}{}_\alpha a^\beta{}_{\mu'} a^{\varepsilon'}{}_\tau T^\alpha{}_\beta{}^\tau. \tag{5.4.2}$$

Consider a physical quantity described by a certain number of components, known through measurements made by a procedure defined in a particular reference frame. Use of the same procedure relative to a second reference frame gives the components of the same physical quantity in the second reference frame. The quantity is represented by a tensor if and only if the components in the two reference frames are related by equations like (5.4.2), whatever the second reference frame. *The equations (5.4.2) characterise the components of a tensor.*

We note that the expressions (5.3.8) allow us to regard $g_{\alpha\beta}$ as the components of a tensor of $T_4^* \otimes T_4^*$. This tensor is called the **metric tensor**. Similarly, from (5.3.9) the $g^{\alpha\beta}$ are the components of a tensor of $T_4 \otimes T_4$.

The number of (non-dummy) indices of the components of a tensor is called the **order** or rank of the tensor. A vector is a tensor of order 1.

Given a tensor **A**, of order n, we obtain a tensor of order $n - 2$, by contracting a pair of indices; for example $A^{\alpha\beta}{}_{\tau\alpha} = B^{\beta}{}_{\tau}$.

The products of the components of two tensors **A** and **B** of order n and m constitute the components of a tensor of order $n + m$; for example $A^{\alpha\beta} \cdot B_{\tau\mu}{}^{\sigma} = T^{\alpha\beta}{}_{\tau\mu}{}^{\sigma}$.

We call **scalar** any quantity invariant under a change of basis, for example the mass of a particle; a scalar is a tensor of order zero.

> As an exercise one can demonstrate the above properties by use of the formulae for changes of basis and transformation of the components of a tensor. One can also verify that the following are scalars:
>   (1) the proper time $d\tau$ between two neighbouring events with coordinates $\{x^{\alpha}\}$ and $\{x^{\alpha} + dx^{\alpha}\}(c^2 d\tau^2 = g_{\alpha\beta} dx^{\alpha} dx^{\beta})$,
>   (2) the scalar product of two vectors,
>   (3) the trace of a tensor of order $n = 2$.

The result of any measurement is a scalar. The numerical value of the result does not depend on which reference frame we use to describe the measurement although we must be very precise about what we are measuring. As an example we consider an electromagnetic wave whose photons have four-momentum $\vec{P}$. An observer stationary in the Minkowski reference frame R measures the frequency v of the wave. Described in R, this measurement has the result $v = cP^0/h$, where P^0 is the time component of $\vec{P}$ in R and h is Planck's constant. In R, the four-vector, $\vec{U}$, of the observer has components $\{c, 0, 0, 0\}$. Hence $v = (\vec{U}, \vec{P})/h$. The same measurement described in the Minkowski reference frame R' leads to the same result. The components of $\vec{U}$ and $\vec{P}$ are changed but the value of $(\vec{U}, \vec{P})/h$ stays the same: this is the frequency measured by the observer. We must clearly distinguish between this measurement, *described* in R', and a measurement of the frequency by an observer stationary in R'. The result of the second measurement is $v' = cP^{0'}/h$ ($v' \neq v$ if R and R' have a non-zero relative motion).

Table 5.1

Definitions

$$\vec{A} \in T_4, \quad \mathbf{A} \in T_4^*:$$

$$A \leftrightarrow \mathbf{A} \Leftrightarrow \mathbf{A}[\vec{V}] = (\vec{A}, \vec{V}) \quad \text{for all } \vec{V} \in T_4$$

$$\{\vec{e}_\alpha\} = \text{basis of } T_4 \qquad \{\mathbf{e}^\alpha\} = \text{dual basis of } T_4^*$$

Fundamental characteristics

Linearity

$$\vec{A} + \mu \cdot \vec{B} \leftrightarrow \mathbf{A} + \mu \cdot \mathbf{B} \text{ for all real } \mu$$

Basis vectors

$$\vec{e}_\alpha \leftrightarrow \mathbf{e}_\alpha = g_{\alpha\beta} \cdot \mathbf{e}^\beta \Leftrightarrow \mathbf{e}^\alpha \leftrightarrow \vec{e}^\alpha = g^{\alpha\beta} \cdot \vec{e}_\beta$$

Contravariant vector $\longleftrightarrow$ covariant vector

$$V^\alpha \cdot (\vec{e}_\alpha) \leftrightarrow V^\alpha \cdot (g_{\alpha\beta} \cdot \mathbf{e}^\beta) = V_\beta \cdot \mathbf{e}^\beta$$

$$V_\alpha = g_{\alpha\beta} V^\beta \Leftrightarrow V^\alpha = g^{\alpha\beta} V_\beta$$

Indices are lowered or raised by contraction with $g_{\alpha\beta}$ or $g^{\alpha\beta}$; for example:

$$T^\alpha{}_\beta \ldots = g_{\beta\sigma} T^{\alpha\sigma} \ldots, \quad T^\alpha{}_\beta \ldots = g^{\alpha\sigma} T_{\sigma\beta} \ldots$$

5.4.4 Covariant and contravariant components of a tensor

The one–one correspondence (cf. § 5.2.3) between vectors of T_4 and T_4^* specified by the metric tensor can be extended to any tensors of the same order. A tensor product of vectors of T_4 (or T_4^*) corresponds to the tensor product obtained by replacing the required number of vectors of T_4 (or T_4^*) by the corresponding vectors of T_4^* (or T_4); for example

$$\vec{A} \otimes \vec{B} \otimes \vec{C} \leftrightarrow \mathbf{A} \otimes \vec{B} \otimes \vec{C} \leftrightarrow \vec{A} \otimes \mathbf{B} \otimes \vec{C} \leftrightarrow \vec{A} \otimes \vec{B} \otimes \mathbf{C}$$

$$\leftrightarrow \mathbf{A} \otimes \mathbf{B} \otimes \vec{C} \leftrightarrow \mathbf{A} \otimes \vec{B} \otimes \mathbf{C} \leftrightarrow \vec{A} \otimes \mathbf{B} \otimes \mathbf{C}$$

$$\leftrightarrow \mathbf{A} \otimes \mathbf{B} \otimes \mathbf{C}.$$

This correspondence is linear, as can easily be verified using the definition above and the linearity of tensor products:

$$\vec{A}_1 \otimes \vec{B} \otimes \vec{C} + \vec{A}_2 \otimes \vec{B} \otimes \vec{C} \leftrightarrow \mathbf{A}_1 \otimes \vec{B} \otimes \vec{C} + \mathbf{A}_2 \otimes \vec{B} \otimes \vec{C}$$

$$\mu \cdot \vec{A} \otimes \vec{B} \otimes \vec{C} \leftrightarrow \mu \cdot \mathbf{A} \otimes \vec{B} \otimes \vec{C}.$$

The correspondence between tensors of the same order is thus completely determined if we know the correspondence between T_4 and $T_4{}^*$. We summarise the properties in table 5.1.

The components of corresponding tensors can be obtained from each other by contraction with $g_{\alpha\beta}$ to "lower" an index, or $g^{\alpha\beta}$ to "raise" an index; we take for example tensors of $T_4 \otimes T_4{}^*$ and those of $T_4{}^* \otimes T_4$:

$$T^\alpha{}_\beta \cdot \vec{e}_\alpha \otimes \mathbf{e}^\beta \leftrightarrow T^\alpha{}_\beta \cdot \mathbf{e}_\alpha \otimes \vec{e}^\beta = T_\alpha{}^\beta \cdot \mathbf{e}^\alpha \otimes \vec{e}_\beta$$

with
$$T^\alpha{}_\beta = g^{\alpha\sigma} \cdot g_{\beta\tau} \cdot T_\sigma{}^\tau \leftrightarrow T_\alpha{}^\beta = g_{\alpha\sigma} \cdot g^{\beta\tau} \cdot T^\sigma{}_\tau.$$

This one–one correspondence means that we can use without ambiguity the expressions "completely contravariant component" of a tensor, when all the indices are raised, and "completely covariant component" when all the indices are lowered; in other cases, the component is called "mixed".

Thus $\delta^\alpha{}_\beta$ and $\delta_\alpha{}^\beta$ are the mixed forms of the metric tensor, while $g^{\alpha\beta}$ is the completely contravariant form.

The above relations between the components of corresponding tensors use the expression for $g_{\alpha\beta}$ and $g^{\alpha\beta}$ in the basis considered; however, these relations define correspondences between tensors which are independent of the basis.

When changing bases, we can raise or lower indices and contract one or more pairs of indices, in any order, before or after changing basis, without changing the result in any way.

As an exercise we suggest that the reader verify these properties, using the formulae for a change of basis given above.

5.5 Classical fields

5.5.1 Generalities

In a Minkowski reference frame, we can associate with every event a quantity [T] which may have several components. For example in a moving fluid, at $\{x, y, z\}$, and at time t, a fluid element has a velocity $\{V^k\}$. To the event $\{x, y, z, t\}$ we may therefore associate the four-component quantity, $[\mathbf{U}] = \{c/\sqrt{1-\beta^2}, \ V^k/\sqrt{1-\beta^2}\}$, with $\beta^2 = \vec{V}^2/c^2$. The components of [U] are determined by measurements whose implementation is defined with respect to the reference frame. In another Minkowski frame we can determine $[\mathbf{U}'] = \{c/\sqrt{1-\beta'^2}, \ V'^k/\sqrt{1-\beta'^2}\}$ by using the same procedure with respect to the new reference frame. We can also measure the relative velocity of the two frames and verify that the components of [U] transform as those of a four-vector. We may then define the vector $\vec{U}$ of T_4 whose components in the natural basis of the first reference frame are [U]. The transformation law from [U] into [U'] assures us that in the second reference frame, "measuring"

[U'] determines the components of the same vector $\vec{U}$ in the new natural basis. We recall that this vector is the four-velocity (§ 1.4 and § 2.3).

With each event E of the space-time, we thus associate a four-vector. The mapping $E \longrightarrow \vec{U}(E)$ constitutes **a field of four-vectors**. This is represented by the four functions $U^\alpha(x, y, z, t)$, the components of $\vec{U}$ at the point with coordinates $\{x, y, z, t\}$, in the natural basis associated with the coordinate system.

Similarly we can introduce scalar fields and more generally fields of tensors.

Because of the fundamental correspondence we need not distinguish between different types of tensor of the same order.

It is convenient to:

(1) relate T_4 to the natural basis associated with the coordinate system describing events,

(2) use for T_4^* the basis dual to that of T_4,

(3) use as basis of $\otimes^n T_4 \otimes^m T_4^*$ the tensor product of basis vectors of T_4 and T_4^*.

Unless otherwise explicitly stated these conventions will always be used. Thus, "a change of coordinates" means "a change of coordinates **and** a change of basis" in such a way that these conventions are maintained.

Let us make a change of coordinates. An event E, with coordinates $x = \{x^\alpha\}$, has the new coordinates $x' = \{x^{\alpha'}\}$.

A field of tensors $\mathbf{T}(E)$ is characterised by its new components regarded as functions of x'. The formulae for changes of coordinates (5.3.11) give the new coordinates following the rule in § 5.4.3; for example for a scalar and tensor of rank three

$$\begin{cases} f'(x') = f(x) \\ T^{\alpha'\beta'}{}_{\mu'}(x') = a^{\alpha'}{}_\sigma a^{\beta'}{}_\tau a^\varepsilon{}_{\mu'} T^{\sigma\tau}{}_\varepsilon(x). \end{cases} \tag{5.5.1}$$

5.5.2 Partial derivatives

Let $f(x)$ be a scalar function, and $f'(x') = f(x)$ the function defining the same scalar field in a new system of coordinates. The chain rule leads to the relations

$$\partial_\alpha f(x) = a^{\beta'}{}_\alpha \partial_{\beta'} f'(x') \quad \text{and} \quad \partial_{\alpha'} f'(x') = a^\beta{}_{\alpha'} \partial_\beta f(x).$$

The transformation rules for partial derivatives of f are those for covariant components of a vector.

Now consider a tensor field, of rank three for example: $\mathbf{T} = T^{\alpha\beta}{}_\tau(x) \cdot \vec{e}_\alpha \otimes \vec{e}_\beta \otimes \mathbf{e}^\tau$. We make an infinitesimal variation of the coordinates, dx^α, corresponding to an infinitesimal displacement in the space-time.

The tensor $\mathbf{T}$ varies infinitesimally: $\mathbf{T}(x + dx) - \mathbf{T}(x) = d\mathbf{T}$.

$$d\mathbf{T} = d\{T^{\alpha\beta}{}_\tau(x)\} \cdot \vec{e}_\alpha \otimes \vec{e}_\beta \otimes \mathbf{e}^\tau = \{T^{\alpha\beta}{}_\tau(x)\}_{,\mu} \cdot dx^\mu \cdot \vec{e}_\alpha \otimes \vec{e}_\beta \otimes \mathbf{e}^\tau. \tag{5.5.2}$$

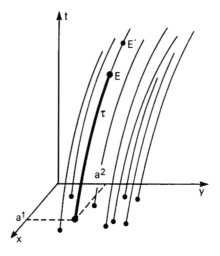

Fig. 5.1

Comparing the rules for changes of coordinates we see that

$$\{T^{\alpha\beta}{}_{\tau}(x)\}_{,\mu} \equiv \partial_{\mu}\{T^{\alpha\beta}{}_{\tau}(x)\} \equiv \frac{\partial\{T^{\alpha\beta}{}_{\tau}(x)\}}{\partial x^{\mu}}$$

represents the components of a tensor of rank four.

We can also verify the following properties:

$$\partial_{\alpha}A^{\alpha} \text{ is a scalar.}$$

$$\partial_{\alpha}\partial^{\alpha}[f(x)] = \partial_{\alpha'}\partial^{\alpha'}[f'(x')] \text{ with } \partial^{\alpha} = g^{\alpha\beta}\partial_{\alpha}.$$

5.5.3 Lagrangian derivative of a tensor field

We consider here a fluid of particles. To each event on the worldline of each particle we associate a tensor with components $T^{\alpha\beta}$, for example.

To describe this flow, there are two possibilities:

(1) The representation introduced above regards $T^{\alpha\beta}$ as a function of the coordinates of the events. This is the **Eulerian representation** of the fluid: $T^{\alpha\beta} = T^{\alpha\beta}(x)$. By differentiation we can define a new tensor field with components $T^{\alpha\beta}(x)_{,\sigma}$: an extra covariant index has appeared.

(2) We could also describe the particles of the fluid by a family of indices; for example the space coordinates, $a = \{a^k\}$, of the particle at the time $t = 0$ (fig. 5.1). For a given particle, $\{a^k\}$ stays fixed. Using the proper time τ as a parameter, the motion of the particle a is described by the functions $x^{\mu} = X^{\mu}(\tau, a)$. This type of representation is called a **Lagrangian representation** of the motion.

The tensor $T^{\alpha\beta}$ is then regarded as a function of τ: $T^{\alpha\beta} = T^{\alpha\beta}(\tau, a)$. Passing from the event E to the neighbouring event E' on the worldline of the same particle, $T^{\alpha\beta}$ varies by $dT^{\alpha\beta}$. We define the Lagrangian derivative of $T^{\alpha\beta}$ along the worldline of the particle as $dT^{\alpha\beta}/d\tau$. *These are the components of a tensor of the same order as* $T^{\alpha\beta}$.

We can re-express the Lagrangian derivative by noting that between E and E', the coordinates have varied by $dx^{\alpha} = U^{\alpha}d\tau$, where $\{U^{\alpha}\}$ is the four-velocity. Introducing the subscripts $_{eu}$ and $_{\ell a}$ to denote the Eulerian and Lagrangian representations at each event E, we get

$$\left\{ \frac{dT^{\alpha\beta}_{(\ell a)}}{d\tau} \right\}_E = \{ U^{\sigma} T^{\alpha\beta}_{(eu),\sigma} \}_E. \tag{5.5.3}$$

5.5.4 Derivative along a vector field

Quite generally, given a vector field with components $A^{\alpha}(x)$, and a tensor field $T^{\alpha\beta\cdots}(x)$, we define the derivative $T^{\alpha\beta\cdots}$ along A^{α} by the relation

$$D_A T^{\alpha\beta\cdots} = A^{\mu} T^{\alpha\beta\cdots}{}_{,\mu}. \tag{5.5.4}$$

We note that these terms are the components of a tensor of the same rank as $T^{\alpha\beta\cdots}$.

The Lagrangian derivative of $T^{\alpha\beta\cdots}$ is the derivative $T^{\alpha\beta\cdots}$ along the four-velocity vector field of the fluid; it gives the variation of the quantity $T^{\alpha\beta\cdots}$ for an observer who follows a particle of the fluid as a function of its proper time.

Chapter 6

The dynamics of continuous media

It is desirable that the reader should be familiar with the fundamentals of the Newtonian theory of continuous media before reading this chapter.

In chapter 4 we considered the relativistic dynamics of a point mass. In this chapter we consider the more general case of continuous media and the description of the electromagnetic field in special relativity. Our aim is to show how the Newtonian laws which hold in the low-velocity limit are generalised in relativity.

6.1 Current density

6.1.1 Particle density

Consider a worldtube in which the particles are distinguished by the index a (cf. § 5.5.3). We consider the space E_t at the instant t in the Minkowski reference frame R. The intersection of the worldtube and E_t constitutes the volume occupied by the fluid at the instant t (fig. 6.1).

Consider an infinitesimal tube, and write dN_a for the number of particles in this tube whose indices lie between a^k and $a^k + da^k$. Let E be any event. The particles occupy at E, within E_t, the volume d^3x. At E, the particle density in R is by definition $\rho_{(P)}$ such that

$$\rho_{(P)}d^3x = dN_a \qquad (6.1.1)$$

with $d^3x = dx\,dy\,dz$.

Let R'_T be the Minkowski reference frame tangent to the worldline of the particle at E. We write $\rho^0_{(P)}$ for the density at E within R'_T; $\rho_{(P)}$ and $\rho^0_{(P)}$ depend on the coordinates of the event E; ρ also depends on the reference frame R, while $\rho^0_{(P)}$ is independent of it; it is a scalar.

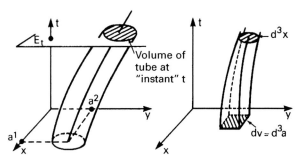

Fig. 6.1

6.1.2 Currents, conservation of particle number

In the reference frame R the fluid flow is described by the velocity field V^k. We define the **current density** of particles at E:

$$j^k \equiv \rho_{(P)}V^k,$$

where all quantities are functions of the coordinates of the event E.

The conservation of particle number can then be written as

$$\frac{\partial \rho_{(P)}}{\partial t} + \partial_k(\rho_{(P)}V^k) = 0. \tag{6.1.2}$$

The situation is formally analogous to the corresponding one in Newtonian theory, so this relation can be proved in the same way.

We define a four-current at each event E, through the relation

$$J_{(P)}{}^\alpha \equiv \rho^0_{(P)}U^\alpha \tag{6.1.3}$$

where U^α is the four-velocity, which is a function of the coordinates of the event E; $J_{(P)}{}^\alpha$ is a four-vector since $\rho^0{}_{(P)}$ is a scalar and U^α is a four-vector.

We can thus show that

$$\partial_\alpha J_{(P)}{}^\alpha = 0 \tag{6.1.4}$$

with

$$\rho_{(P)} = \rho^0_{(P)}\frac{U^0}{c} = \frac{J_{(P)}{}^0}{c}. \tag{6.1.5}$$

To prove (6.1.4) and (6.1.5), we consider an elementary tube of matter and its tangent reference frame R'_T at E and the reference frame R derived from it by the Lorentz transformation

$$t = \frac{t' + Vx'/c^2}{\sqrt{1-\beta^2}}, \quad x = \frac{x' + Vt'}{\sqrt{1-\beta^2}}, \quad y = y', \quad z = z'.$$

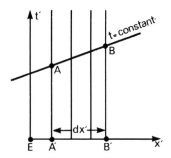

The matter velocity at E with respect to R is that of R'_T with respect to R, i.e. $\{V, 0, 0\}$.

Eliminating t', we find $x = x'\sqrt{(1 - \beta^2)} + Vt$. Thus for the events A and B of the above figure with the same coordinate t

$$dx \equiv x_B - x_A = (x'_B - x'_A)\sqrt{(1 - \beta^2)}.$$

Near E the matter tube is comoving with R'_T. In the space associated with R'_T the volume occupied by the tube is

$$dx'\,dy'\,dz' = d^3x',$$

with $dx' \equiv x'_{B'} - x'_{A'} = x'_B - x'_A$.

In the space associated with R, the volume occupied by the matter tube is $dx\,dy\,dz = d^3x$, with $dx = x_B - x_A = dx'\sqrt{(1 - \beta^2)}$, $dy = dy'$ and $dz = dz'$.

We thus get $d^3x = d^3x'\sqrt{(1 - \beta^2)}$. We note in passing that d^3x is smaller than d^3x' (cf. § 3.2.4).

The number of particles in the matter tube is $dN_a = \rho_{(P)}d^3x = \rho^0{}_{(P)}d^3x'$, so that $\rho_{(P)} = \rho^0{}_{(P)}/\sqrt{(1 - \beta^2)}$. We write $\rho_{(P)} = \rho^0{}_{(P)}U^0/c$, where U^α is the four-velocity of matter at E. The latter expression is general if R and R' are related by any Poincaré transformation. We note the relation

$$\rho_{(P)}\{c, V^k\} = \rho^0{}_{(P)}\{U^0, U^0V^k\} = \rho^0{}_{(P)}\{U^\alpha\} = \{J_{(P)}{}^\alpha\}.$$

This satisfies the relations (6.1.4) and (6.1.5) if we use (6.1.2).

If there are nuclear reactions in the fluid the conservation laws apply to those particles whose numbers are conserved in these reactions.

Finally, if all the particles have the same mass m and charge q, we define the mass current, $J_{(m)}{}^\alpha = mJ_{(P)}{}^\alpha$, and the charge current, $J^\alpha = qJ_{(P)}{}^\alpha$, which both satisfy conservation relations like (6.1.4).

6.2 The electromagnetic field tensor

In a Minkowski reference frame R, the electromagnetic field is characterised by the electric and magnetic "vectors" $\{E^k\}$ and $\{B^k\}$. In reality these are not vectors but the components of an antisymmetric rank 2 tensor $F^{\alpha\beta} = -F^{\beta\alpha}$, **the electromagnetic field tensor**. One of the elegant features of the theory of relativity is the combining of the electric and magnetic fields into a single tensor.

$$F^{0k} = \sqrt{\varepsilon_0}\, E^k$$

$$F^{12} = \frac{B^3}{\sqrt{\mu_0}}, \qquad F^{23} = \frac{B^1}{\sqrt{\mu_0}}, \qquad F^{31} = \frac{B^2}{\sqrt{\mu_0}}$$

where ε_0 and μ_0 are the dielectric permittivity and magnetic permeability of the vacuum ($\varepsilon_0\mu_0 c^2 = 1$). The antisymmetry of the tensor fixes the other components.

We recall that Maxwell's equations have the "vectorial" form

$$\mathrm{div}\left[\overline{B}\right] = 0 \qquad\qquad \mathrm{curl}\left[\overline{E}\right] = -\frac{\partial\overline{B}}{\partial t}$$

$$\mathrm{div}\left[\overline{E}\right] = \frac{\rho}{\varepsilon_0} \qquad\qquad \mathrm{curl}\left[\overline{B}\right] = \mu_0\left(\overline{j} + \varepsilon_0\frac{\partial\overline{E}}{\partial t}\right)$$

where ρ is the charge density in the reference frame considered, $\rho = J^0/c$, and $\overline{j}$ is the current density, $j^k = \rho V^k = J^k$; $\{J^\alpha\}$ is the charge current defined in the preceding paragraph.

We can verify that Maxwell's equations may be written as:

$$\begin{cases} F_{\alpha\beta,\tau} + F_{\tau\alpha,\beta} + F_{\beta\tau,\alpha} = 0 \\ (F^{\alpha\beta})_{,\beta} = \sqrt{\mu_0}\, J^\alpha. \end{cases} \qquad (6.2.1)$$

Relations like these, making manifest the tensorial nature of each side, are called "covariant forms" of the Maxwell equations.

In its tangent reference frame R'_T a particle is static. The only force acting upon it is therefore electric: $f^{k'} = qE^{k'}$. We have seen above (§ 4.1) that the four-force in the proper rest-frame is expressed as $F^{0'} = 0$, $F^{k'} = f^{k'}$, which we can synthesise as $F^{\beta'} = (q\sqrt{\mu_0})U_{\alpha'}F^{\alpha'\beta'}$. This relation expresses the equality of two four-vectors in R'_T. In an arbitrary reference frame the four-force can therefore be written as

$$F^\beta = (q\sqrt{\mu_0})U_\alpha F^{\alpha\beta}. \qquad (6.2.2)$$

As an exercise the reader may verify that the expression for the relativistic force is that given in § 4.4.3 and show using (6.2.2) that the antisymmetry of $F^{\alpha\beta}$ implies the relation $F^\alpha U_\alpha = 0$.

6.3 Energy–momentum tensor

6.3.1 Non-interacting fluid

Consider a fluid whose particles do not interact with each other. Each particle feels the external force F^α. Multiplying each side of (4.1.3) by $\rho^0{}_{(P)}$, we get

$$\Phi^\alpha = \rho^0{}_{(m)}\frac{dU^\alpha}{d\tau} = \rho^0{}_{(m)}\frac{dx^\beta}{d\tau}\partial_\beta U^\alpha = \rho^0{}_{(m)}U^\beta(U^\alpha)_{,\beta}$$

where Φ^α is the **four-force density**.

The conservation law for matter currents is:

$$\partial_\alpha J_{(m)}{}^\alpha \equiv (\rho^0{}_{(m)}U^\alpha)_{,\alpha} = 0.$$

These laws can be grouped together by introducing the **energy–momentum tensor**, $T^{\alpha\beta}$:

◆
$$T^{\alpha\beta} \equiv \rho^0{}_{(m)}U^\alpha U^\beta.$$

The equation of motion of the fluid and the conservation of mass can thus be written as:

◆
$$\partial_\alpha T^{\alpha\beta} = \Phi^\beta.$$

We must however retain the constraint satisfied by the four-force density:

$$\Phi^\alpha U_\alpha = 0.$$

We form $U_\beta T^{\alpha\beta}{}_{,\alpha}$, with $T^{\alpha\beta}{}_{,\alpha} = U^\beta(\rho^0{}_{(m)}U^\alpha)_{,\alpha} + (U^\beta)_{,\alpha}(\rho^0{}_{(m)}U^\alpha) = \Phi^\beta$. We use the relation $U^\alpha U_\alpha = c^2$ and the relation $U_\beta U^\beta{}_{,\alpha} = 0$ which follows from it, so that $U_\beta\Phi^\beta = 0$. We thus obtain the conservation law $(\rho^0{}_{(m)}U^\alpha)_{,\alpha} = 0$. Using this law, as well as the formula $U^\alpha(\ )_{,\alpha} \equiv d(\ )/d\tau$, we find the equation of motion.

6.3.2 Continuous media

The Newtonian theory of continuous media is governed by two equations:
 – Mass conservation (continuity):

$$\frac{\partial\rho}{dt} + \partial_k(\rho V^k) = 0.$$

 – The equation of motion:

$$\rho\frac{\partial V^k}{\partial t} + \rho V^j\partial_j V^k = \sigma^{jk}{}_{,j}$$

where $\sigma^{jk} = \sigma^{kj}$ is the three-dimensional stress tensor, V^k is the Eulerian velocity field of the matter particles, and ρ is the mass density; all these quantities are functions of both position and time.

To find the relativistic generalisation of these laws, we consider the tangent reference frame, R'_T, to the matter worldline at E, an arbitrary event of M_4. In this frame, we identify $\sigma^{j'k'}$ with the Newtonian stress tensor and define $\sigma^{0'\alpha'} = \sigma^{\alpha'0'} = 0$. Then the Newtonian laws are

$$\partial_{\alpha'} T^{\alpha'\beta'} = 0$$

where $T^{\alpha'\beta'} \equiv \rho^0_{(m)} U^{\alpha'} U^{\beta'} - \sigma^{\alpha'\beta'}$ and $U^{\alpha'} = \{c, V^{k'}\}$.

Thus for $\beta' = 0$ we get at E, $\partial_{\alpha'} T^{\alpha'0'} = c\{\partial\rho^0/\partial t' + \partial_{k'}(\rho^0 V^{k'})\} = 0$: this is the conservation of mass in the Newtonian approximation. For $\beta' = k'$ we get $\partial_{\alpha'} T^{\alpha'k'} = 0$ with $\partial_{\alpha'} T^{\alpha'k'} = \partial_{0'}(\rho^0 c V^{k'}) + \partial_{j'}(\rho^0 V^{j'} V^{k'}) - \partial_{j'}\sigma^{j'k'}$, where $\sigma^{j'k'}$ is the Newtonian stress tensor. Using the conservation law demonstrated above, we recover the equation of motion.

We thus introduce the tensor $\sigma^{\alpha\beta}$ whose components in R'_T are $\sigma^{\alpha'\beta'}$. The equations of motion are:

$$\partial_\alpha T^{\alpha\beta} = 0 \tag{6.3.1}$$

where $T^{\alpha\beta}$ is a symmetrical tensor, the **energy–momentum tensor**

$$T^{\alpha\beta} = \rho^0_{(m)} U^\alpha U^\beta - \sigma^{\alpha\beta} \quad \text{with} \quad U_\alpha \sigma^{\alpha\beta} = 0. \tag{6.3.2}$$

The latter relation can easily be proved in the tangent reference frame: it proves that the vector with components $U_\alpha \sigma^{\alpha\beta}$ in R vanishes.

In the presence of an external four-force density Φ^β, the equations of motion are

$$\partial_\alpha T^{\alpha\beta} = \Phi^\beta. \tag{6.3.3}$$

The energy–momentum tensor of a perfect fluid is

$$T^{\alpha\beta} = \left(\rho^0 + \frac{P}{c^2}\right) U^\alpha U^\beta - P\eta^{\alpha\beta} \tag{6.3.4}$$

where P is the pressure and $\rho^0 c^2$ the energy density in the tangent reference frame.

We note that the current $\rho^0 U^\alpha$ is not in general conserved.

6.3.3 The interpretation of $T^{0\beta}$

For a non-interacting fluid made of particles of mass m it is clear that T^{0k}/c is the momentum density and that T^{00} is the energy density. The momentum of an elementary tube of particles can be written

$$dP^\alpha = dN_a m U^\alpha = \rho_{(P)} d^3x m U^\alpha = \rho^0_{(P)} \frac{U^0}{c} d^3x m U^\alpha = \frac{T^{0\alpha}}{c} d^3x.$$

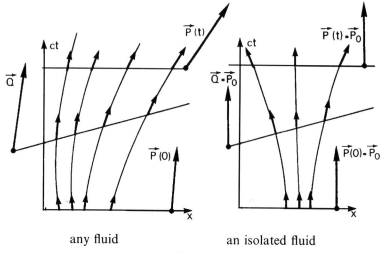

any fluid   an isolated fluid

Fig. 6.2

In the general case, we define the momentum density as $T^{0\alpha}/c$. Then the total energy–momentum P^α satisfies the relation

$$cP^\alpha = \iiint_t T^{0\alpha} d^3x \qquad (6.3.5)$$

where the integral extends over the spatial section of the matter tube. We note that P^α agrees with the definition given in § 4.2 for a system of particles, constituting a non-interacting fluid.

The four-vector $\vec{P}(t)$ with components P^α depends on the reference frame R and the time t.

In another reference frame the same definition gives the four-momentum $\vec{Q}(t')$. However, for an isolated system, $\Phi^\alpha = 0$, and if it has finite spatial extent, $T^{\alpha\beta}(x) = 0$ for $x^2 + y^2 + z^2 > A(t)$, we can show that $\vec{P}$ is constant with $\vec{P} = \vec{Q}$ (fig. 6.2).

$$\frac{dP^\alpha}{dt} = 0. \qquad (6.3.6)$$

Given the coordinate t, we integrate over a volume V enclosing the matter:

$$\frac{dP^\alpha}{dt} = \iiint_V \frac{\partial T^{0\alpha}}{c\partial t} d^3x = \iiint_V \partial_0 T^{0\alpha} d^3x = \iiint_V -\partial_k(T^{k\alpha}) d^3x$$

The last integral can be transformed into an integral over the surface Σ bounding the volume V by use of Gauss's theorem:

$$\iiint_V -\partial_k(T^{k\alpha}) d^3x = \iint_\Sigma (T^{k\alpha}) n_k d\Sigma$$

where $\{n^k\}$ is the vector normal to Σ, $n^k = -n_k$, and $d\Sigma$ is the element of area on Σ. $T^{k\alpha}$ vanishes on Σ, and we deduce that P^α is constant.

To show that $\vec{P} = \vec{Q}$ we consider a special Lorentz transformation taking us between the Minkowski reference frames R and R':

$$x = \frac{x' - \beta ct'}{\sqrt{(1-\beta^2)}}, \quad y = y', \quad z = z', \quad t = \frac{t' - \beta x'/c}{\sqrt{(1-\beta^2)}}.$$

The matrix for the change of basis is written $a^{\alpha'}{}_\beta$.

We define $Q^{\alpha'}$ in R', following (6.3.5):

$$cQ^{\alpha'} = \iiint_{t'} T^{0'\alpha'} d^3 x'$$

with $T^{0'\alpha'} = a^{0'}{}_\sigma a^{\alpha'}{}_\beta T^{\sigma\beta}$.

Performing this integral with $t' = $ constant we see that it is independent of t', as $Q^{\alpha'}$ is constant like P^α. We set $t' = 0$.

We introduce the components Q^α of $\vec{Q}$ in R, $Q^\alpha = a^\alpha{}_{\beta'} Q^{\beta'}$:

$$cQ^\alpha = a^\alpha{}_{\beta'} a^{0'}{}_\sigma a^{\beta'}{}_\tau \iiint_{t'=0} T^{\sigma\tau} d^3 x'.$$

In this expression, $T^{\sigma\tau}$ is regarded as a function of t' and x'. We show that $Q^\alpha = P^\alpha$ for an isolated system.

We use the relation $a^\alpha{}_{\beta'} a^{\beta'}{}_\tau = \delta^\alpha{}_\tau$ and write out explicitly the coefficients $a^{0'}{}_\sigma$:

$$cQ^\alpha = \iiint_{t'=0} \frac{1}{\sqrt{1-\beta^2}} \{T^{0\alpha} + \beta T^{1\alpha}\} d^3 x'.$$

To perform the integration we use the change of variables defined by the Lorentz transformation for $t' = 0$:

$$t = -\frac{\beta x}{c}; \quad x = \frac{x'}{\sqrt{1-\beta^2}}.$$

The quantities $T^{\alpha\sigma}(x, y, z, t,)$ are functions of x, y, z alone:

$$T^{\alpha\sigma} = T^{\alpha\sigma}\left(x, y, z, -\frac{\beta x}{c}\right).$$

From the change of variable we have $dx\sqrt{1-\beta^2} = dx'$, $dy = dy'$, $dz = dz'$; so that

$$cQ^\alpha = \iiint_V \{T^{0\alpha} + \beta T^{1\alpha}\} d^3 x.$$

Q^α thus appears as a function of β:

$$c\frac{dQ^\alpha}{d\beta} = \iiint_V \left\{\frac{dT^{0\alpha}}{d\beta} + T^{1\alpha} + \beta\frac{dT^{1\alpha}}{d\beta}\right\} d^3 x$$

with $\dfrac{dT^{\alpha\sigma}}{d\beta} = -x\partial_0 T^{\alpha\sigma} \Rightarrow \dfrac{dT^{0\alpha}}{d\beta} = -x\partial_0 T^{0\alpha} = x(\partial_1 T^{1\alpha} + \partial_2 T^{2\alpha} + \partial_3 T^{3\alpha})$, where ∂_1 means "derivative wrt the first variable".

Thus $\dfrac{\partial T^{\alpha\sigma}}{\partial x} = \partial_1 T^{\alpha\sigma} - \beta\partial_0 T^{\alpha\sigma}.$

Using these relations, we find

$$c\frac{dQ^\alpha}{d\beta} = \iiint_V \left\{ \frac{\partial(xT^{1\alpha})}{\partial x} + \frac{\partial(xT^{2\alpha})}{\partial y} + \frac{\partial(xT^{3\alpha})}{\partial z} \right\} d^3x.$$

For fixed index α Gauss's theorem can be used to transform the volume integral into a surface integral:

$$c\frac{dQ^\alpha}{d\beta} = -\iint_\Sigma \left(xT^{k\alpha}\right) n_k d\Sigma \quad \text{where } \Sigma \text{ is the surface bounding } V.$$

This quantity vanishes since $T^{k\alpha}$ is zero on Σ. Hence Q^α is independent of β. Q^α can be calculated for $\beta = 0$; we find $Q^\alpha = P^\alpha$, as we wished to show.

6.3.4 Energy–momentum tensor of the electromagnetic field

Consider a charged medium, subject to an electromagnetic field with tensor $F^{\alpha\beta}$. The four-force density, Φ^α, can be found from the four-force (6.2.2) and the equations of motion are $\partial_\alpha T^{\alpha\beta} = \sqrt{\mu_0} J_\alpha F^{\alpha\beta}$. Using Maxwell's equations (6.2.1) to eliminate J_α, we find $\partial_\alpha T^{\alpha\beta} = (F_\alpha{}^\tau)_{,\tau} F^{\alpha\beta}$. We can show that these equations can be put in the form

$$\partial_\tau \left\{ (T^{\tau\beta}) + \left(-F_\alpha{}^\tau F^{\alpha\beta} + \frac{1}{4}\eta^{\tau\beta} F_{\alpha\sigma} F^{\alpha\sigma} \right) \right\} = 0.$$

In fact $(F_\alpha{}^\tau)_{,\tau} F^{\alpha\beta} = (F_\alpha{}^\tau F^{\alpha\beta})_{,\tau} - F_\alpha{}^\tau (F^{\alpha\beta})_{,\tau}$. Using the first set of Maxwell equations, we find $F_\alpha{}^\tau (F^{\alpha\beta})_{,\tau} = F_{\alpha\tau} F^{\alpha\beta,\tau} = -F_{\alpha\tau} F^{\tau\alpha,\beta} - F_{\alpha\tau} F^{\beta\tau,\alpha}$.

Changing the dummy indices and using the antisymmetry of $F_{\alpha\beta}$, we find $F_\alpha{}^\tau (F^{\alpha\beta})_{,\tau} = -F_{\alpha\tau} F^{\tau\alpha,\beta} - F_\alpha{}^\tau (F^{\alpha\beta})_{,\tau}$.

Further $F_\alpha{}^\tau (F^{\alpha\beta})_{,\tau} = -\dfrac{1}{2} F_{\alpha\tau} F^{\tau\alpha,\beta} = -\dfrac{1}{4} (F_{\alpha\tau} F^{\tau\alpha})^{,\beta}.$

Thus

$$(F_\alpha{}^\tau)_{,\tau} F^{\alpha\beta} = (F_\alpha{}^\tau F^{\alpha\beta})_{,\tau} + \frac{1}{4}(F_{\alpha\tau} F^{\tau\alpha})^{,\beta} = \left\{ (F_\alpha{}^\tau F^{\alpha\beta}) - \frac{1}{4}\eta^{\tau\beta}(F_{\alpha\sigma} F^{\alpha\sigma}) \right\}_{,\tau}.$$

The equations thus have the claimed form.

The system {medium + electromagnetic field} is an isolated system whose energy–momentum tensor we can define as

$$T_{(\text{tot})}{}^{\alpha\beta} = T^{\alpha\beta} + \Theta^{\alpha\beta}$$

where $\Theta^{\alpha\beta}$ is called the "energy–momentum tensor for the electromagnetic field":

$$\blacklozenge \qquad \Theta^{\tau\beta} = -F_\alpha{}^\tau F^{\alpha\beta} + \frac{1}{4}\eta^{\tau\beta} F_{\alpha\sigma} F^{\alpha\sigma} \qquad\qquad (6.3.7)$$

We note that this tensor is symmetric.

The equations of motion can then be written

$$\partial_\alpha T^{\alpha\beta}_{(\text{tot})} = 0. \tag{6.3.8}$$

♦

The components $\Theta^{0\alpha}$ are easily found:

$$\Theta^{00} = \frac{\varepsilon_0}{2}\overline{E}^2 + \frac{1}{2\mu_0}\overline{B}^2 \qquad \Theta^{0k} = \Theta^{k0} = -\frac{\sqrt{\varepsilon_0}}{\sqrt{\mu_0}}(\overline{E} \wedge \overline{B})^k.$$

Θ^{00} and Θ^{0k}/c are respectively the energy density and momentum density of the electromagnetic field.

Consider a charged medium of finite spatial extent for all t. The radiated electromagnetic power is the flux of the Poynting pseudo-vector $-\{c\Theta^{0k}\} = \overline{E} \wedge \overline{B}/\mu_0$ over any surface enclosing the matter distribution.

The total energy in the volume V is

$$E = \iiint_V T^{00}_{(\text{tot})} d^3x.$$

Just as before use of the conservation equation and Gauss's theorem gives

$$\frac{dE}{dt} = -\iiint_V c\partial_k T_{(\text{tot})}{}^{0k} d^3x = \iint_\Sigma c\Theta^{0k} n_k d\Sigma$$

where the volume V is bounded by the surface Σ assumed to enclose the matter distribution.

Under these conditions we have global conservation laws for the total momentum and the total energy of the system {matter distribution + electromagnetic field}.

This conservation property is very general and, extended to any type of field which binds composite particles, is precisely the property we used in the study of collisions.

In the absence of charges, $J^\alpha = 0$, Maxwell's equations imply $\partial_\alpha \Theta^{\alpha\beta} = 0$. The medium and the electromagnetic field thus satisfy independent equations: the two systems are decoupled.

Chapter 7

General coordinate systems

Minkowski coordinates allow us to specify events in M_4. Their advantages include the simple expression for proper time, the simple form of the equations of motion for a free particle, and above all the easy use of the Newtonian limit for finding the relativistic forms of physical laws and interpreting the results. Nevertheless there are cases where Minkowski coordinates do not match the physical situation well and it is preferable for example to use a reference frame comoving with the measuring apparatus.

We have already considered a generalisation of Minkowski coordinates in chapter 5, where we considered systems derived from them by relations of the form (5.3.11). We shall complete this generalisation here by introducing general coordinates in M_4 and the corresponding natural bases of T_4. We shall not attempt to make precise the way the coordinates are determined physically, but try to give a general formulation of the preceding results. *The expressions we shall find are formally valid for the class of theories known as "metric theories of gravitation", which in particular include general relativity.*

In this chapter we shall introduce quantities of the same type as those introduced earlier, particularly in chapter 5. However, these quantities are here purely "local", i.e. they depend on the coordinates of an event E, chosen *a priori* but otherwise arbitrary.

The index E indicating that various quantities depend on the event considered is not repeated systematically. Similarly the argument $x = \{x_E{}^\alpha\}$ is usually omitted from functions which depend on it. We shall usually make these omissions if there is no risk of ambiguity: the reader should pay attention to this possibility in this chapter.

7.1 General coordinate systems and the natural associated basis

7.1.1 The natural reference frame

Let $\{x^\alpha\}$ be a Minkowski coordinate system in M_4 and $\{\vec{e}_\alpha\}$ the natural associated basis of T_4.

We make the following coordinate change:

$$x^{\alpha'} = F^{\alpha'}(x^\beta). \tag{7.1.1}$$

The functions $F^{\alpha'}$ are not completely arbitrary as the old coordinates must be expressible as functions of the new ones in the region where the new system is used. This must hold for the events to be unambiguously specified:

$$x^\beta = G^\beta(x^{\alpha'}). \tag{7.1.2}$$

Consider the event E with coordinates $\{x_E{}^\alpha\}$ and $\{x_E{}^{\alpha'}\}$ in the two systems. Near E, expanding (7.1.1) to first order gives

$$x^{\alpha'} \approx x_E{}^{\alpha'} + a^{\alpha'}{}_\beta(x_E)(x^\beta - x_E{}^\beta) \tag{7.1.3}$$

where we have set

$$a^{\alpha'}{}_\beta \equiv \frac{\partial F^{\alpha'}}{\partial x^\beta}.$$

We can invert (7.1.3) by use of the inverse, $a^\tau{}_{\alpha'}$, of $a^{\alpha'}{}_\beta$:

$$a^\tau{}_{\alpha'}a^{\alpha'}{}_\beta = \delta^\tau{}_\beta \qquad a^{\alpha'}{}_\tau a^\tau{}_{\beta'} = \delta^{\alpha'}{}_{\beta'}. \tag{7.1.4}$$

Identifying the resulting expression with the expansion of (7.1.2), we find

$$x^\alpha \approx G^\alpha(x'_E) + a^\alpha{}_{\beta'}(x'_E)(x^{\beta'} - x_E{}^{\beta'})$$

with

$$a^\tau{}_{\alpha'} = \frac{\partial G^\tau}{\partial x^{\alpha'}}.$$

The expression (7.1.3), regarded as exact, defines a new coordinate system, $\{x^{\alpha''}\}$, which coincides with $\{x^{\alpha'}\}$ at E. The passage from $\{x^\alpha\}$ to $\{x^{\alpha''}\}$ is of the same type as the coordinate change studied in § 5.3.4. The natural basis of this new system is called the "natural basis" for the coordinate system $\{x^{\alpha'}\}$ at E. It is written $\{\dot{e}_{\alpha'}\}$:

$$(\dot{e}_{\alpha'})_E = (a^\beta{}_{\alpha'})_E \cdot \dot{e}_\beta. \tag{7.1.5}$$

The "natural reference system" at E is $\{E, \dot{e}_{\alpha'}\}$. In practice we often use the terms "natural basis" and "natural reference system" interchangeably; both depend on the coordinate system and the event E considered.

7.1.2 Interpretation

Consider the space-time curves along which only one coordinate varies: $x^{\alpha'} = $ const for all α' other than the given value $\underline{\alpha}'$. Such a curve $\mathscr{C}_{\underline{\alpha}'}$ is called a **coordinate line**.

The coordinate line $\mathscr{C}_{1'}$ of the new system passing through the event E with coordinates $x_E{}^{\alpha'}$ satisfies the following parametric equation, given in the old coordinate system as: $x^\alpha = G^\alpha(x_E{}^{0'}, u, x_E{}^{2'}, x_E{}^{3'}) = X^\alpha(u)$, where $x^{1'}$ is chosen as parameter: $x^{1'} = u$.

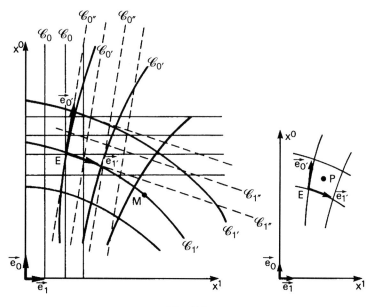

Fig. 7.1

Let M be a point of $\mathscr{C}_{1'}$ and $\overrightarrow{EM}$ the corresponding vector of T_4;
$\overrightarrow{EM} = (X^\alpha(u) - X^\alpha(u^0)) \cdot \vec{e}_\alpha$, where $X^\alpha(u^0) = x_E{}^\alpha$:

$$\left(\frac{d\overrightarrow{EM}}{du}\right)_E = \left(\frac{dX^\alpha}{du}\right)_E \cdot \vec{e}_\alpha = \left(\frac{\partial G^\alpha}{\partial x^{1'}}\right)_E \cdot \vec{e}_\alpha = (a^\alpha{}_{1'})_E \cdot \vec{e}_\alpha = (\vec{e}_{1'})_E.$$

The vector $d\overrightarrow{EM}/du = \vec{e}_{1'}$ is tangent at E to the coordinate line $\mathscr{C}_{1'}$. The coordinate lines of the various systems we have introduced are shown in fig. 7.1.

Consider an arbitrary event P, with coordinates $\{x^\alpha\}$. The vector $\overrightarrow{EP} = (x^\alpha - x_E{}^\alpha) \cdot \vec{e}_\alpha$ can be regarded as a function of the coordinates $x^{\alpha'}$ of the point P in the new coordinate system. Quite generally we can verify the relations

$$\blacklozenge \qquad (\vec{e}_{\alpha'})_E = \left(\frac{\partial \overrightarrow{EP}}{\partial x^{\alpha'}}\right)_E = \left(\frac{\partial \overrightarrow{\Omega P}}{\partial x^{\alpha'}}\right)_E \qquad (7.1.6)$$

where Ω is an arbitrary event of M_4 chosen once and for all.

This expression defines the natural basis $\{\vec{e}_{\alpha'}\}$ in an intrinsic way, without reference to a pre-chosen Minkowski basis.

7.1.3 General changes of coordinates

We consider *two arbitrary coordinate systems* $\{x^\alpha\}$ and $\{x^{\alpha'}\}$. The coordinates x^α can be written as functions of the $x^{\beta'}$ and conversely $x^{\alpha'}$ can be written as functions of the x^β. We set

$$\blacklozenge \qquad a^{\alpha'}{}_\beta = \frac{\partial x^{\alpha'}}{\partial x^\beta} \quad \text{and} \quad a^\alpha{}_{\beta'} = \frac{\partial x^\alpha}{\partial x^{\beta'}} \qquad (7.1.7)$$

where $a^{\alpha'}{}_\beta$ and $a^\alpha{}_{\beta'}$ are functions of the coordinates.

As above, we can establish the relations

$$\blacklozenge \qquad a^\tau{}_{\alpha'} \cdot a^{\alpha'}{}_\beta = \delta^\tau{}_\beta \quad \text{and} \quad a^{\alpha'}{}_\tau \cdot a^\tau{}_{\beta'} = \delta^{\alpha'}{}_{\beta'} \qquad (7.1.8)$$

as well as the relations analogous to (7.1.5) between the natural basis vectors associated with each coordinate system at each point:

$$\blacklozenge \qquad \vec{e}_\alpha = a^{\mu'}{}_\alpha \cdot \vec{e}_{\mu'}, \quad \vec{e}_{\mu'} = a^\alpha{}_{\mu'} \cdot \vec{e}_\alpha. \qquad (7.1.9)$$

We emphasise that the various quantities we have introduced, $a^{\alpha'}{}_\beta$, $\vec{e}_\alpha$, etc., depend on the space-time point considered; thus $\vec{e}_\alpha$ is a vector field.

> To prove the above properties we introduce an additional Minkowski reference frame. We express $\vec{e}_\alpha$ and $\vec{e}_{\alpha'}$ as functions of basis vectors of this reference frame. Using the definitions (7.1.6) and (7.1.7) and the chain rule gives (7.1.9).

7.1.4 Tensor fields in local coordinates

Given a tensor field, we can associate a tensor with each event E, of coordinates $x_E = \{x_E{}^\alpha\}$. We decompose this tensor in the natural basis at E. Consider a tensor field **T** defined by its components $T^\alpha{}_\beta(x)$ in the natural basis $\{\vec{e}_\alpha\}$ associated with an arbitrary coordinate system $\{x^\alpha\}$:

$$(\mathbf{T})_E = T^\alpha{}_\beta(x_E) \cdot (\vec{e}_\alpha \otimes \mathbf{e}^\beta)_E.$$

The quantities $T^\alpha{}_\beta(x)$ are the components of the tensor field in **local coordinates**; they are functions of the coordinates x^β of the event considered.

We note that variations of $T^\alpha{}_\beta$ from one point to another have two distinct origins: the variation of **T** and the variation of the basis $\{\vec{e}_\alpha\}$. We study these variations and the associated derivatives in the next section.

We decompose all tensors in local coordinates. Thus the metric tensor has local coordinate components

$$g_{\alpha\beta} = g_{\beta\alpha} = (\vec{e}_\alpha, \vec{e}_\beta)$$

where $g_{\alpha\beta}$ is a function of the coordinates $x = \{x^\alpha\}$ of the event E considered, as $\vec{e}_\alpha$ depends on the event E.

All the local expressions for scalar products, movement of indices, and changes of basis at the space-time point E remain formally unchanged from those established in chapter 5. However, the quantities $a^\alpha{}_{\beta'}$, $a^{\alpha'}{}_\beta$, $g_{\alpha\beta}$, $g^{\alpha\beta}$, etc. here depend on the event E: they are functions of the coordinates. In particular the line element is written

$$\blacklozenge \qquad ds^2 = c^2 d\tau^2 = g_{\alpha\beta}(x) dx^\alpha dx^\beta. \qquad (7.1.10)$$

Table 7.1

N.B. The E specifying that quantities depend on the event considered is not repeated systematically. Similarly the argument $x = \{x_E{}^\alpha\}$ is frequently omitted from functions depending on it. These omissions are the general rule if there is no ambiguity.

$\{x^\alpha\}$ and $\{x^{\alpha'}\}$ are two arbitrary coordinate systems of M_4.
$\{\vec{e}_\alpha\}_E$ and $\{\vec{e}_{\alpha'}\}_E$ are the natural associated basis fields of T_4.
$\{\mathbf{e}^\alpha\}_E$ and $\{\mathbf{e}^{\alpha'}\}_E$ are the bases of T_4^* dual to $\{\vec{e}_\alpha\}_E$ and $\{\vec{e}_{\alpha'}\}_E$

$$\mathbf{e}^\alpha[\vec{e}_\beta] = \delta^\alpha{}_\beta \qquad \mathbf{e}^{\alpha'}[\vec{e}_{\beta'}] = \delta^{\alpha'}{}_{\beta'}$$

The metric tensor has components $g_{\alpha\beta}(x)$

$$g_{\alpha\beta}(x) \equiv (\vec{e}_\alpha, \vec{e}_\beta)_E$$
$$g_{\alpha\tau}g^{\tau\beta} \equiv \delta_\alpha{}^\beta \equiv \delta^\beta{}_\alpha = g^{\beta\tau}g_{\tau\alpha}$$
$$g_{\alpha\beta} = g_{\beta\alpha} \qquad\qquad g^{\alpha\beta} = g^{\beta\alpha}$$

The line element is

$$ds^2 = c^2 d\tau^2 = g_{\alpha\beta}(x)dx^\alpha dx^\beta$$

Raising and lowering of indices is performed using $g_{\alpha\beta}$ and $g^{\alpha\beta}$

$$T^\alpha{}_\beta \ldots (x) = g^{\alpha\sigma}(x)g_{\beta\mu}(x)T_\sigma{}^\mu \ldots (x)$$

The scalar product of two vectors of T_4, $\vec{V}_E$ and $\vec{W}_E$, is

$$(\vec{V}, \vec{W})_E = g_{\alpha\beta}(x)V^\alpha(x)W^\beta(x) = V^\alpha(x)W_\alpha(x) = V_\alpha(x)W^\alpha(x)$$

Coordinate changes

$$a^\alpha{}_{\beta'}(x) \equiv \left(\frac{\partial x^\alpha}{\partial x^{\beta'}}\right)_E \qquad a^{\alpha'}{}_\beta(x) \equiv \left(\frac{\partial x^{\alpha'}}{\partial x^\beta}\right)_E$$
$$a^\alpha{}_{\tau'}(x)a^{\tau'}{}_\beta(x) = \delta^\alpha{}_\beta \qquad a^{\alpha'}{}_\tau(x)a^\tau{}_{\beta'}(x) = \delta^{\alpha'}{}_{\beta'}$$
$$(\vec{e}_\alpha)_E = a^{\beta'}{}_\alpha(x)(\vec{e}_{\beta'})_E \qquad \vec{e}_{\alpha'} = a^\beta{}_{\alpha'}\vec{e}_\beta$$
$$(\mathbf{e}^\alpha)_E = a^\alpha{}_{\beta'}(x)(\mathbf{e}^{\beta'})_E \qquad \mathbf{e}^{\alpha'} = a^{\alpha'}{}_\beta\mathbf{e}^\beta$$
$$T^\alpha{}_\beta{}^\mu(x) = a^\alpha{}_{\tau'}(x)a^{\sigma'}{}_\beta(x)a^\mu{}_{\varepsilon'}(x)T^{\tau'}{}_{\sigma'}{}^{\varepsilon'}(x)$$
$$T^{\alpha'}{}_{\beta'}{}^{\mu'} = a^{\alpha'}{}_\tau a^\sigma{}_{\beta'}a^{\mu'}{}_\varepsilon T^\tau{}_\sigma{}^\varepsilon$$

Table 7.1 continued

Connection $\Gamma^\alpha{}_{\beta\mu}$, **and Christoffel symbols** $\{{}^\alpha{}_{\beta\mu}\}$

$$\Gamma^\alpha{}_{\beta\mu}(x) = \Gamma^\alpha{}_{\mu\beta}(x) = \{{}^\alpha{}_{\beta\mu}\}$$

$$\{{}^\alpha{}_{\beta\mu}\} \equiv \frac{1}{2} g^{\alpha\sigma} \{g_{\sigma\beta,\,\mu} + g_{\sigma\mu,\,\beta} - g_{\beta\mu,\sigma}\}$$

Variations of the basis vectors

$$d\vec{e}_\alpha \equiv \Gamma^\beta{}_{\alpha\mu} dx^\mu \vec{e}_\beta \qquad d\mathbf{e}^\alpha \equiv -\Gamma^\alpha{}_{\beta\mu} dx^\mu \mathbf{e}^\beta$$

Variation of a tensor

$$\mathbf{T}_E \equiv T^{\alpha\beta}{}_\tau(x)(\vec{e}_\alpha \otimes \vec{e}_\beta \otimes \mathbf{e}^\tau)_E$$
$$d\mathbf{T} \equiv \nabla_\mu T^{\alpha\beta}{}_\tau(x)(\vec{e}_\alpha \otimes \vec{e}_\beta \otimes \mathbf{e}^\tau)_E dx^\mu$$

Covariant derivative

$$\nabla_\mu T^{\alpha\beta\cdots}{}_{\tau\cdots} \equiv (T^{\alpha\beta\cdots}{}_{\tau\cdots})_{;\mu} \qquad \partial_\mu T^{\alpha\beta\cdots}{}_{\tau\cdots} \equiv (T^{\alpha\beta\cdots}{}_{\tau\cdots})_{,\mu} \equiv \frac{\partial T^{\alpha\beta\cdots}{}_{\tau\cdots}}{\partial x^\mu}$$

$$\nabla_\mu T^{\alpha\beta}{}_\tau = (T^{\alpha\beta}{}_\tau)_{,\mu} + \Gamma^\alpha{}_{\sigma\mu} T^{\sigma\beta}{}_\tau + \Gamma^\beta{}_{\sigma\mu} T^{\alpha\sigma}{}_\tau - \Gamma^\sigma{}_{\tau\mu} T^{\alpha\beta}{}_\sigma$$

Dynamics of a point mass m

$$\frac{dU^\alpha}{d\tau} + \Gamma^\alpha{}_{\beta\mu} U^\beta U^\mu = \frac{F^\alpha}{m}$$

Dynamics of continuous media

Energy–momentum tensor: $T^{\alpha\beta} = T^{\beta\alpha}$
Equations of motion: $\nabla_\alpha T^{\alpha\beta} = 0$

We summarise these and other formulae in table 7.1.

We illustrate the introduction of local coordinates by an example from plane geometry in two dimensions.

Consider an orthonormal plane Cartesian reference frame $\{O, \vec{i}, \vec{j}\}$; the coordinates are written $\{x, y\}$ and we use the notation $\vec{e}_x = \vec{i}, \vec{e}_y = \vec{j}$. The metric tensor has components g_{jk}: $g_{kk} = 1$ and $g_{jk} = 0$ for $j \neq k$, with $j, k, \ldots = 1$ or 2.

Consider the polar coordinate system defined by the coordinate changes

$$x = r \cos\theta \quad y = r \sin\theta.$$

It is instructive to use this example in all the preceding machinery. In particular we can determine $a^\alpha{}_{\beta'}$:

$$a^x{}_r = \cos\theta, \quad a^x{}_\theta = -r \sin\theta, \quad a^y{}_r = \sin\theta, \quad a^y{}_\theta = r \cos\theta.$$

The natural frame at the point with coordinates $\{r, \theta\}$ is

$$\vec{e}_r = a^x{}_r \cdot \vec{e}_x + a^y{}_r \cdot \vec{e}_y = \cos\,\theta \cdot \vec{i} + \sin\,\theta \cdot \vec{j}$$
$$\vec{e}_\theta = a^x{}_\theta \cdot \vec{e}_x + a^y{}_\theta \cdot \vec{e}_y = -r\,\sin\,\theta \cdot \vec{i} + r\,\cos\,\theta \cdot \vec{j}$$

We can determine the components of the metric tensor in the new frame. Thus we can verify that the distance, dL, between two neighbouring points of coordinates $\{r, \theta\}$ and $\{r + dr, \theta + d\theta\}$ can be written as

$$dL^2 = dx^2 + dy^2 = g_{kj}dx^k dx^j = dr^2 + r^2 d\theta^2.$$

We can draw the coordinate lines and verify that at each point the vectors of the natural associated basis are tangent to them.

7.2 Small variations and derivatives

To study the variation of a tensor, we have to take into account both the variation of its components and the variation of the basis vectors as we move in the space-time.

7.2.1 The linear connection in T_4

Consider two neighbouring events A and B of M_4, with coordinates $\{x^\alpha\}$ and $\{x^\alpha + dx^\alpha\}$ respectively, where $\{x^\alpha\}$ is an arbitrary coordinate system and $\{\vec{e}_\alpha\}$ the natural associated basis.

We set $d\vec{e}_\alpha = (\vec{e}_\alpha)_B - (\vec{e}_\alpha)_A$. The vector $d\vec{e}_\alpha$ can be expressed with respect to the basis $\{(\vec{e}_\alpha)_A\}$; the following results are first order in dx^μ, and remain unchanged if we use the natural basis of B. We therefore omit the subscript $_A$ in what follows.

The decomposition can be written

♦
$$d\vec{e}_\alpha \equiv \Gamma^\beta{}_{\alpha\mu} dx^\mu \cdot \vec{e}_\beta \qquad (7.2.1)$$

with

$$\Gamma^\beta{}_{\alpha\mu} \cdot \vec{e}_\beta = \frac{\partial \vec{e}_\alpha}{\partial x^\mu}.$$

The coefficients $\Gamma^\beta{}_{\alpha\mu}$ are functions of x^α, called **linear connection coefficients**. Relation (7.1.6) implies

$$\Gamma^\beta{}_{\alpha\mu} \cdot \vec{e}_\beta = \frac{\partial \vec{e}_\alpha}{\partial x^\mu} = \frac{\partial^2 \overrightarrow{\Omega P}}{\partial x^\mu \partial x^\alpha} = \frac{\partial^2 \overrightarrow{\Omega P}}{\partial x^\alpha \partial x^\mu} = \Gamma^\beta{}_{\mu\alpha} \cdot \vec{e}_\beta$$

so the connection is symmetric:

♦
$$\Gamma^\beta{}_{\alpha\mu} = \Gamma^\beta{}_{\mu\alpha}. \qquad (7.2.2)$$

The connection coefficients can be expressed as functions of the metric tensor:

$$\Gamma^\beta{}_{\alpha\mu} = \frac{1}{2}g^{\beta\tau}(g_{\tau\alpha,\mu} + g_{\tau\mu,\alpha} - g_{\alpha\mu,\tau}). \tag{7.2.3}$$

We call the quantity $\{^\beta{}_{\alpha\mu}\}$ the **Christoffel symbol**:

$$\{^\beta{}_{\alpha\mu}\} \equiv \frac{1}{2}g^{\beta\tau}(g_{\tau\alpha,\mu} + g_{\tau\mu,\alpha} - g_{\alpha\mu,\tau}). \tag{7.2.4}$$

We also introduce the notation

$$\Gamma_{\alpha\,\beta\mu} \equiv g_{\alpha\tau} \cdot \Gamma^\tau{}_{\beta\mu} = \frac{1}{2}(g_{\alpha\beta,\mu} + g_{\alpha\mu,\beta} - g_{\beta\mu,\alpha}) = \Gamma_{\alpha\,\mu\beta}$$

$$\{_{\alpha\,\beta\mu}\} \equiv g_{\alpha\tau} \cdot \{^\tau{}_{\beta\mu}\} \equiv \frac{1}{2}(g_{\alpha\beta,\mu} + g_{\alpha\mu,\beta} - g_{\beta\mu,\alpha}) \equiv \{_{\alpha\,\mu\beta}\}.$$

In special relativity the connection coefficients are equal to the Christoffel symbols.

The proof is as follows. Using the properties of the scalar product, we find $dg_{\alpha\beta} = (d\vec{e}_\alpha, \vec{e}_\beta) + (\vec{e}_\alpha, d\vec{e}_\beta)$; from the defining relation for $\Gamma^\beta{}_{\alpha\mu}$, we get

$$g_{\alpha\beta,\mu} = (\Gamma^\sigma{}_{\alpha\mu}\vec{e}_\sigma, \vec{e}_\beta) + (\vec{e}_\alpha, \Gamma^\tau{}_{\beta\mu}\vec{e}_\tau) = g_{\sigma\beta}\Gamma^\sigma{}_{\alpha\mu} + g_{\alpha\tau}\Gamma^\tau{}_{\beta\mu}$$

so that $g_{\alpha\beta,\mu} = \Gamma_{\beta\,\alpha\mu} + \Gamma_{\alpha\,\beta\mu}$. This relation and those deduced from it by permuting indices give

$$g_{\alpha\beta,\mu} + g_{\mu\alpha,\beta} - g_{\beta\mu,\alpha} = \Gamma_{\beta\,\alpha\mu} + \Gamma_{\alpha\,\beta\mu} + \Gamma_{\alpha\,\mu\beta} + \Gamma_{\mu\,\alpha\beta} - \Gamma_{\mu\,\beta\alpha} - \Gamma_{\beta\,\mu\alpha}.$$

Using symmetry and raising indices we thus find the relation (7.2.3).

To study the transformation rules for the connection coefficients we consider a second arbitrary coordinate system $\{x^{\alpha'}\}$. As before in (7.1.7), we introduce $a^{\alpha'}{}_\beta$ and $a^\alpha{}_{\beta'}$. We can thus show the relations

$$\Gamma^{\varepsilon'}{}_{\beta'\mu'} = a^\tau{}_{\mu'}\{a^{\varepsilon'}{}_\sigma a^\alpha{}_{\beta'}\Gamma^\sigma{}_{\alpha\tau} + a^{\varepsilon'}{}_\alpha a^\alpha{}_{\beta',\tau}\}. \tag{7.2.5}$$

In either coordinate system we have

$$\partial_\mu\vec{e}_\beta = \Gamma^\alpha{}_{\beta\mu}(x)\vec{e}_\alpha(x) \qquad \partial_{\mu'}\vec{e}_{\beta'} = \Gamma^{\varepsilon'}{}_{\beta'\mu'}(x')\vec{e}_{\varepsilon'}(x').$$

We use the relations

$$\vec{e}_{\beta'} = a^\alpha{}_{\beta'}\vec{e}_\alpha \quad \text{and} \quad \partial_{\mu'} = a^\tau{}_{\mu'}\partial_\tau.$$

We thus find

$$\partial_{\mu'}\vec{e}_{\beta'} = \Gamma^{\varepsilon'}{}_{\beta'\mu'}\vec{e}_{\varepsilon'} = a^\tau{}_{\mu'}\partial_\tau(a^\alpha{}_{\beta'}\vec{e}_\alpha) = a^\tau{}_{\mu'}(a^\alpha{}_{\beta',\tau}\vec{e}_\alpha + a^\alpha{}_{\beta'}\Gamma^\sigma{}_{\alpha\tau}\vec{e}_\sigma).$$

Expressing $\vec{e}_\alpha$ as a function of $\vec{e}_{\alpha'}$ and equating the coefficients of the basis vectors $\vec{e}_{\alpha'}$, we get (7.2.5).

7.2.2 The linear connection in T_4^*

Consider natural bases $\{\vec{e}_\alpha\}$ and the associated dual bases $\{\mathbf{e}^\alpha\}$. We can define the connection $\Gamma^{*\alpha}{}_{\beta\mu}$ of T_4^* through the relation

$$\blacklozenge \qquad\qquad d\mathbf{e}^\alpha = \Gamma^{*\alpha}{}_{\tau\mu} dx^\mu \cdot \mathbf{e}^\tau \qquad\qquad (7.2.6)$$

where $d\mathbf{e}^\alpha$ is the variation of $\mathbf{e}^\alpha$ between two neighbouring events.

The connections of T_4 and T_4^* are not independent as the bases in each of the spaces remain dual to each other when we move in M_4. We can show that

$$\blacklozenge \qquad\qquad \Gamma^{*\alpha}{}_{\tau\mu} = -\Gamma^\alpha{}_{\tau\mu}. \qquad\qquad (7.2.7)$$

The definition of dual bases implies

$$(\mathbf{e}^\alpha[\vec{e}_\beta])_A = (\mathbf{e}^\alpha[\vec{e}_\beta])_B = \left((\mathbf{e}^\alpha)_A + d\mathbf{e}^\alpha\right)\left[(\vec{e}_\beta)_A + d\vec{e}_\beta\right] = \delta^\alpha{}_\beta.$$

We omit the subscript A in the remainder of the calculation. Expanding, we get

$$(d\mathbf{e}^\alpha)[\vec{e}_\beta] = -\mathbf{e}^\alpha[d\vec{e}_\beta] = -\Gamma^\tau{}_{\beta\mu}\, dx^\mu\, \delta^\alpha{}_\tau = \Gamma^{*\alpha}{}_{\tau\mu}\, dx^\mu\, \delta^\tau{}_\beta.$$

This leads to (7.2.7).

7.2.3 Variation of a tensor field

Consider the tensor field $\mathbf{T} = T^\alpha{}_\beta \cdot \vec{e}_\alpha \otimes \mathbf{e}^\beta$. The variation of $\mathbf{T}$ between the events A and B is $d\mathbf{T} = \mathbf{T}_B - \mathbf{T}_A$. Using the expressions for $d\vec{e}_\alpha$ and $d\mathbf{e}^\beta$ we find

$$d\mathbf{T} = dT^\alpha{}_\beta \cdot \vec{e}_\alpha \otimes \mathbf{e}^\beta + T^\alpha{}_\beta \cdot d\vec{e}_\alpha \otimes \mathbf{e}^\beta + T^\alpha{}_\beta \cdot \vec{e}_\alpha \otimes d\mathbf{e}^\beta$$
$$= \left\{ T^\alpha{}_{\beta,\mu} \cdot \vec{e}_\alpha \otimes \mathbf{e}^\beta + T^\alpha{}_\beta \Gamma^\tau{}_{\alpha\mu} \cdot \vec{e}_\tau \otimes \mathbf{e}^\beta - T^\alpha{}_\beta \Gamma^\beta{}_{\tau\mu} \cdot \vec{e}_\alpha \otimes \mathbf{e}^\tau \right\} dx^\mu.$$

We thus get the expressions

$$\blacklozenge \qquad \begin{cases} d\mathbf{T} = (\nabla_\mu T^\alpha{}_\beta)dx^\mu \cdot \vec{e}_\alpha \otimes \mathbf{e}^\beta \\ \nabla_\mu T^\alpha{}_\beta \equiv \partial_\mu(T^\alpha{}_\beta) + \Gamma^\alpha{}_{\tau\mu} T^\tau{}_\beta - \Gamma^\tau{}_{\beta\mu} T^\alpha{}_\tau. \end{cases} \qquad (7.2.8)$$

$d\mathbf{T}$ is called the **absolute differential** and $\nabla_\mu T^\alpha{}_\beta$ the **covariant derivative** of the tensor field.

The quantity $d\mathbf{T}$ is a tensor; $\nabla_\mu T^\alpha{}_\beta dx^\mu$ is therefore the component of a tensor whatever the infinitesimal element dx^μ. We deduce that $\nabla_\mu T^\alpha{}_\beta$ *represents the components of a tensor*. We leave the proof as an exercise. In a Minkowski system $\Gamma^\alpha{}_{\beta\mu} = 0$, and the components of the tensor $\nabla_\mu T^\alpha{}_\beta$ reduce to $\partial_\mu T^\alpha{}_\beta$. A constant tensor field is thus characterised by the vanishing of $\nabla_\mu T^\alpha{}_\beta$ in local coordinates. As an exercise the reader should make explicit the reasoning leading from the relation $T^\alpha{}_\beta = $ const, which holds in a Minkowski system, to the vanishing of the covariant derivative.

These expressions can be generalised to tensors having an arbitrary number of indices.

In section 5.5 we defined the derivative of a tensor field along the vector field A^α. In local coordinates this tensor field can be written in the form

$$\blacklozenge \qquad D_A \mathbf{T} = A^\mu \cdot \nabla_\mu T^\alpha{}_\beta \cdot \vec{e}_\alpha \otimes \mathbf{e}^\beta. \qquad (7.2.9)$$

We can further demonstrate the following properties:

$$\nabla_\mu g_{\alpha\beta} = \nabla_\mu g^{\alpha\beta} = 0 \qquad (7.2.10)$$
$$\nabla_\mu(T \cdot R) = \nabla_\mu(T) \cdot (R) + (T) \cdot \nabla_\mu(R)$$

where T and R represent the components of arbitrary tensors of which some indices may be contracted. These properties show that *the covariant derivative of a tensor can be calculated before or after contracting, raising or lowering indices.*

7.3 Dynamical equations

7.3.1 Dynamics of a point mass

In an arbitrary system of coordinates we represent the motion of a point mass by introducing the four functions $x^\alpha = X^\alpha(\tau)$, where τ is the proper time measured along the worldline of the point mass.

The four-velocity has the components

$$U^\alpha(\tau) = dx^\alpha/d\tau \qquad (7.3.1)$$

in local coordinates.

To demonstrate (7.3.1) we show that $\{U^\alpha(\tau)\}$ is a four-vector by finding its transformation under a coordinate change. We then verify that $U^\alpha(\tau)$ is the component of the four-velocity in a Minkowski coordinate system.

In local coordinates the four-acceleration, $\{\Gamma^\alpha(\tau)\}$, is

$$\Gamma^\alpha = \frac{dU^\alpha}{d\tau} + \Gamma^\alpha{}_{\beta\mu} \cdot U^\beta \cdot U^\mu. \qquad (7.3.2)$$

The relation (7.3.2) can be found by considering two neighbouring events on the particle's worldline. The variation of $\vec{U} = U^\alpha \cdot \vec{e}_\alpha$ is given using (7.2.1):

$$d\vec{U} \equiv \nabla U^\alpha \vec{e}_\alpha = dU^\alpha \vec{e}_\alpha + U^\alpha d\vec{e}_\alpha.$$

We thus find $\nabla U^\alpha = dU^\alpha + \Gamma^\alpha{}_{\beta\mu} dx^\mu U^\beta$ where $\Gamma^\alpha{}_{\beta\mu}$ is a function of τ, $\Gamma^\alpha{}_{\beta\mu} = \Gamma^\alpha{}_{\beta\mu}[X^\sigma{}_{(\tau)}]$, and $dx^\mu = U^\mu d\tau$. The definition of $\vec{\Gamma}$, $\vec{\Gamma} d\tau = d\vec{U}$, gives the desired result.

The equation of motion (4.1.4) thus has the form

$$\frac{dU^\alpha}{d\tau} + \Gamma^\alpha{}_{\beta\mu} U^\beta U^\mu = \frac{F^\alpha}{m} \tag{7.3.3}$$

where m is the particle's mass and $\{F^\alpha\}$ are the components of the four-force in local coordinates, while $\Gamma^\alpha{}_{\beta\mu}$ is given by (7.2.3). As the various quantities are evaluated at the specified point on the worldline, these are functions of τ.

> As an exercise, determine the motion of a free particle by using the fact that the proper time is extremal on its worldline. It then follows that the relation (7.2.3) gives (7.3.3) with $F^\alpha = 0$.

7.3.2 Dynamics of continuous media

To write the equations of chapter 6 in local coordinates we replace the ∂_μ symbols by ∇_μ. We justify this procedure by an example.

Just as in chapter 6, we introduce the particle four-current $J_{(P)}{}^\alpha = \rho^0{}_{(P)} U^\alpha$, and (6.1.4) becomes

$$\nabla_\alpha J_{(P)}{}^\alpha = 0. \tag{7.3.4}$$

The left-hand side of (7.3.4) is a scalar. In a Minkowski coordinate system we can express this scalar as $\partial_\alpha J_{(P)}{}^\alpha$ and the equations (7.3.4) and (6.1.4) both express the same property, i.e. that this scalar vanishes.

The rules established in chapter 6 express equalities between tensors. It is convenient therefore to express the equality of their components in local coordinates. Thus we introduce the charge current J^α and the tensor $F^{\alpha\beta}$. Maxwell's equations are then

$$\begin{cases} F_{\alpha\beta;\tau} + F_{\tau\alpha;\beta} + F_{\beta\tau;\alpha} = 0 \\ (F^{\alpha\beta})_{;\beta} = \sqrt{\mu_0} \cdot J^\alpha. \end{cases} \tag{7.3.5}$$

We have used the notation $\nabla_\mu(\ ) = (\ )_{;\mu}$.

Similarly we can find the local coordinate components of the energy–momentum tensors introduced in chapter 6 by substituting $g_{\alpha\beta}(x)$ and $g^{\alpha\beta}(x)$ for $\eta_{\alpha\beta}$ and $\eta^{\alpha\beta}$; thus for a perfect fluid

$$T^{\alpha\beta} = \left(\rho^0 + \frac{P}{c^2}\right) U^\alpha U^\beta - P g^{\alpha\beta} \tag{7.3.6}$$

while for the electromagnetic field

$$\theta^{\tau\beta} = -F_\alpha{}^\tau F^{\alpha\beta} + \frac{1}{4} g^{\tau\beta} F_{\alpha\sigma} F^{\alpha\sigma}. \tag{7.3.7}$$

The dynamical equations (6.3.8) can be written in local coordinates in the form

$$\nabla_\alpha T^{\alpha\beta} \equiv T^{\alpha\beta}{}_{;\alpha} = 0 \tag{7.3.8}$$

where $T^{\alpha\beta}(x)$ stands for the total energy–momentum tensor of the matter and fields present.

7.3.3 Inertial forces and the Newtonian limit

When motions are slow, Newtonian theory is acceptable. In this theory we can regard an accelerated reference frame as Galilean provided we introduce inertial forces. We shall show by an example how the notation above allows us to recover this result in the limit of small velocities.

Consider a Minkowski frame R, $\{O, x, y, t\}$. We "forget" here the z coordinate, as the problem is confined to the plane $z = 0$.

A rotating disc in R rotates about O, with constant angular velocity Ω. More precisely we assume that the matter points constituting the disc have motions characterised as in Newtonian theory for a rigid disc, by $x = x' \cos \Omega t - y' \sin \Omega t$ and $y = x' \sin \Omega t + y' \cos \Omega t$ where x' and y' are constants. We can regard these expressions as coordinate changes and use the reference frame R', $\{O, x', y', t' = t\}$. The new coordinate system is therefore comoving with the disc.

Consider a fixed point in the new system. In Newtonian language the force keeping this point fixed is equal and opposite to the inertial force; in relativistic language this is the four-force $F^{\alpha'}$. The equations of motion (7.3.3) give us $F^{\alpha'}$ as functions of $g_{\alpha'\beta'}$, the components of the metric tensor in R'. We can thus verify that in the Newtonian limit the components $F^{x'}$ and $F^{y'}$ of $F^{\alpha'}$ reduce to the negatives of the components of the inertial force.

Given two neighbouring events, we have

$$dx = dx' \cos \Omega t - dy' \sin \Omega t - \Omega dt(x' \sin \Omega t + y' \cos \Omega t)$$
$$dy = dx' \sin \Omega t + dy' \cos \Omega t + \Omega dt(x' \cos \Omega t - y' \sin \Omega t)$$
$$dt = dt'.$$

The line element is then

$$ds^2 = c^2 d\tau^2 = c^2 dt^2 - dx^2 - dy^2$$
$$= c^2(1 - \Omega^2 (x'^2 + y'^2)/c^2)dt'^2 - 2\Omega dt'(x'dy' - y'dx') - dx'^2 - dy'^2.$$

The metric tensor in R' thus has the form

$$g_{0'0'} = 1 - \Omega^2(x'^2 + y'^2)/c^2,$$

$$g_{0'x'} = g_{x'0'} = \Omega y'/c, \qquad\qquad g_{0'y'} = g_{y'0'} = -\Omega x'/c,$$
$$g_{x'x'} = g_{y'y'} = -1, \qquad\qquad g_{x'y'} = g_{y'x'} = 0.$$

Inverting the matrix of components $g_{\alpha'\beta'}$ gives $g^{\alpha'\beta'}$. The relations (7.2.3) give the connection coefficients.

For the frame considered x' and y' are constants: $U^{x'} = dx'/d\tau = 0$, $U^{y'} = dy'/d\tau = 0$. The relation, $(\vec{U}, \vec{U}) = c^2$ gives U^0.

The equations of motion (7.3.3) here become

$$\Gamma^{x'}{}_{0'0'}\, U^{0'} U^{0'} = F^{x'}/m, \qquad \Gamma^{y'}{}_{0'0'}\, U^{0'} U^{0'} = F^{y'}/m$$

In the limit of small velocities, $\Omega^2(x'^2 + y'^2) \ll c^2$, we find

$$F^{x'} = -m\Omega^2 x' \qquad\qquad F^{y'} = -m\Omega^2 y'.$$

Under these conditions $-F^{x'}$ and $-F^{y'}$ reduce to the components of the Newtonian inertial force, which here is the centrifugal force.

We note that the motion we have assumed for the disc is only acceptable for $\Omega^2(x'^2 + y'^2) < c^2$ as $d\tau^2$ must be positive for a point mass such as a part of the disc. Moreover the hypothesis for the disc's motion is heuristic as we have no theory for the behaviour of the matter of the disc.

This example shows that the components of the metric tensor describe the relativistic equivalent of inertial forces appearing in non-Minkowski reference frames.

7.3.4 Minkowski coordinate systems

We have seen that constructing Galilean reference frames with Newtonian hindsight leads to the construction of Minkowski reference frames (cf. § 2.2.2). We can now show this quite generally.

In the Newtonian framework the construction of a Galilean reference frame requires the following three coordinate conditions:

(1) Clocks always measure the time (1.1.2) for any reference frame with arbitrary velocity V^k.

(2) Free particles satisfy (2.1.1).

(3) The distance between two fixed points satisfies (1.1.1). We still have to specify how distances are measured. If we use a length standard we have to specify its behaviour in special relativity and interpret the measurement made by a Newtonian physicist. We are not equipped to proceed in this way. Thus we replace this condition by another property of orthonormal Galilean reference frames: in an elastic collision of a particle of mass m with another mass M which is initially at rest, with $M \gg m$, the quantity $\sum V^k V^k$ remains unchanged in the collision, where V^k is the particle velocity.

We assume that special relativity is a rigorously exact theory while Newtonian theory is only approximate. The coordinate system is defined by physical operations followed by calculations. In such an operatively-defined coordinate system the three properties above are empirical and can be checked by experiment.

The last two properties are satisfied in both relativity and Newtonian theory provided the coordinate systems are suitably chosen. The Newtonian physicist we consider can verify (2) and (3) whatever the velocities of the particles he studies.

By contrast the first property is not satisfied in relativity. We therefore regard it as false, while the Newtonian physicist believes it to be correct. This comes from the fact that he can test this property to finite accuracy, such that $(V^j V^k/c^2)$ is small compared with unity although V^k/c need not be.

Given that the three properties above are verified under the specified conditions, the Newtonian physicist deduces that the coordinate system is Galilean. The relativistic physicist who knows the "true theory" deduces that the coordinate system is Minkowskian within the accuracy of measurement.

Thus we have shown that coordinate systems found to be Galilean in the limit of small velocities are in reality Minkowski coordinate systems in special relativity.

The metric on the coordinate system considered is $g_{\alpha\beta}$. The first property implies

$$dt^2 = d\tau^2 \Rightarrow 1 = g_{00} + 2g_{0k}\frac{V^k}{c} + g_{jk}\frac{V^j}{c}\frac{V^k}{c}.$$

This relation cannot be rigorously satisfied for any V^k. However, if measurement accuracy is limited and the last term is negligible we find

$$g_{00} = 1 + O(\varepsilon^2) \text{ and } g_{0k} = O(\varepsilon)$$

where ε^2 is a number of the order of magnitude of the best relative accuracy that can be measured: V^k/c thus does not exceed some term of order ε. $O(\varepsilon^n)$ is a quantity of order ε^n.

The four-velocity is written U^α, and we use the relation $U^k = U^0 V^k/c$. To exploit the last condition, we first establish the relation between U^α and $g_{\alpha\beta}$. Then we write the conservation of four-momentum in the collision as

$$c^2 = g_{00}(U^0)^2 + 2g_{0k}U^0 U^k + g_{jk}U^j U^k$$

$$mU^0{}_m + MU^0{}_M = mU^0{}_{m'} + MU^0{}_{M'}$$

$$mU^k{}_m + MU^k{}_M = mU^k{}_{m'} + MU^k{}_{M'}$$

where $U^\alpha{}_m$ and $U^\alpha{}_{m'}$ are the four-velocities of the particle of mass m before and after the collision. The particle of mass M is initially at rest, so that $U^k{}_M = 0$ and $U^0{}_M = c/\sqrt{g_{00}}$.

We calculate $U^k{}_M$ and $U^0{}_M$ using the last two equations. We use the results to express the first relation. After some simplification we get to first order in m/M a relation satisfied for any $V^k{}_m$ and $V^k{}_{m'}$ such that $\sum V^k{}_m V^k{}_m = \sum V^k{}_{m'} V^k{}_{m'} \leqslant c^2$. This condition leads to $g_{ik} = G\delta_{ik} + O(\varepsilon^2)$ where δ_{ik} is the Kronecker symbol and G is a function of the coordinates x^α.

The second condition implies that the equations (7.3.3) are satisfied for $F^k = 0$ and $x^k = V^k t + a^k$ with $x^0 = ct$. We consider the motion of a "photon". The Newtonian physicist applies the above equation with $\sum V^k V^k = c^2$. The relativistic physicist also knows that $d\tau^2$ vanishes along the worldline determined by the Newtonian physicist, which implies $G = -1 + O(\varepsilon^2)$ and $g_{0k} = O(\varepsilon^2)$, if we use the earlier results. Thus we get

$$g_{00} = 1 + O(\varepsilon^2), \quad g_{\alpha\beta} = O(\varepsilon^2) \text{ when } \alpha \neq \beta$$

$$g_{11} = -1 + O(\varepsilon^2), \quad g_{22} = -1 + O(\varepsilon^2), \quad g_{33} = -1 + O(\varepsilon^2).$$

We therefore verify that the equations (7.3.3) with $F^\alpha = 0$ are satisfied to the measurement accuracy by the trajectories of free particles observed by the Newtonian physicist.

II

Relativistic gravitation

Chapter 1

Newtonian gravitation

Familiarity with the material of appendix A is desirable before reading this chapter.

In this chapter we shall use Galilean coordinate systems which are either Cartesian $\{x, y, z, t\}$ or spherical $\{r, \theta, \phi, t\}$.

1.1 The Newtonian equivalence principle

In Newtonian theory gravity is an attractive force between two material particles (fig. 1.1).

The expression for the force exerted by the particle P_1 on the particle P_2 is

$$\vec{F} = G M_1 M_2 \frac{\vec{u}}{r^2}$$

with $r = \|\overrightarrow{P_1 P_2}\|$, $\vec{u} = \dfrac{\overrightarrow{P_2 P_1}}{r}$; M_1 and M_2 are coefficients characterising the particles, called the **gravitational masses**, which play the role of gravitational "charges" and are usually expressed in kilograms; G is the **Newtonian constant of gravitation**:

$$G = 6.67 \times 10^{-11} \mathrm{m}^3 \mathrm{s}^{-2} \mathrm{kg}^{-1}.$$

Consider a given distribution of matter and let us place a point mass P of gravitational mass M at $\vec{x} = \{x, y, z\}$. The gravitational force acting on this mass is

$$\vec{F} = M \cdot \vec{g} \quad \text{with} \quad \vec{g} = \vec{g}(\vec{x}) = G \cdot \sum_a M_a \frac{\vec{x}_a - \vec{x}}{\|\vec{x}_a - \vec{x}\|^3} \tag{1.1.1}$$

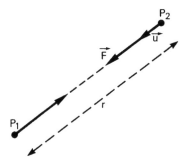

Fig. 1.1

where $\sum_a$ implies that the summation is extended over all points of the matter distribution (denoted a) other than P; this is Newton's law of universal gravitation.

$\vec{g}$ is called the **gravitational field** at the point with coordinates $\{x, y, z\}$.

When acted on by the force $\overrightarrow{F}$, the particle P experiences acceleration $\vec{\gamma}$ given by

$$\overrightarrow{F} = m \cdot \vec{\gamma}.$$

where m is a coefficient characterising the inertia of P; m is called the **inertial mass**. We have seen in special relativity that mc^2 is the energy of the point mass P.

It is experimentally well verified (to an accuracy exceeding one part in 10^{12} in some cases) that

$$m = M$$

with both masses measured in SI units. This relation therefore implies that the acceleration of the particle P is independent of its mass: $\vec{\gamma} = \vec{g}$.

The equality of gravitational and inertial masses constitutes the **Newtonian equivalence principle**.

Thus, in an artificial satellite, all objects feel the same gravitational acceleration towards the Earth, even though they have different masses. The identity of their free-fall motions in the terrestrial gravitational field, which is a consequence of the Newtonian equivalence principle, implies that objects are apparently weightless within the satellite.

However, these objects also experience their mutual gravitational attractions as well as the terrestrial field. In an artificial satellite these are negligible; but this is not always the case. Although the Earth is a satellite of the Sun, its gravitational attraction is not negligible in its neighbourhood, as it is responsible for the fact that bodies fall towards it and the existence of Earth satellites.

The Newtonian equivalence principle is the basis for the relativistic metric theories of gravity presented in the following chapter. Experimental demonstrations of this principle are thus particularly important. Already by the end of the 17th century Newton had verified the equality of gravitational and inertial masses to an accuracy η of order 10^{-3}. To understand the idea of his experiment, we can study the motion of a simple pendulum while distinguishing between the two types of mass. The equality of the two masses implies that the period of the pendulum depends only on its length. This allows an experimental check of the principle. In the 19th century Eötvös used a different method to obtain $\eta \approx 5 \times 10^{-9}$. Currently the most accurate experiments are those of Dicke in Princeton in 1964 ($\eta \approx 10^{-11}$) and of Braginskii in Moscow in 1974 ($\eta \approx 10^{-12}$). Measurements of satellites should allow one to reach the value of $\eta \approx 3 \times 10^{-17}$ to within two orders of magnitude, implying that η could be between 10 and 100 times larger or smaller.

The Newtonian equivalence principle can be analysed more closely by postulating only that the gravitational and inertial masses are proportional: $M = km$. The law of gravitational attraction is then $F = KMM'/r^2 = Kk^2mm'/r^2$, where K is a universal constant. In the case considered k is independent of the body, its position in space and time, so we set $G = Kk^2$. In this form the principle is called the "strong equivalence principle". It is, however, possible to imagine that k does not depend on the body but does depend on its position and on time. In this case the equivalence principle is called "weak".

It would also be possible to distinguish between two kinds of gravitational masses: active and passive. Given two particles, the active mass of the first, M_A, creates the gravitational field acting on the passive mass, M'_P, of the second: $F = KM_AM'_P/r^2$.

As an exercise the reader should find the relations assuring the equality of action and reaction and discuss what would happen to the centre of mass of an isolated system of two particles if this equality fails.

In the present context the reader could also specify the various equivalence principles compatible with the equality of action and reaction, or with the identical motion of two purely passive particles in the gravitational field of a central mass, given the same initial conditions.

The relation (1.1.1) allows us to express $\vec{g}$ as the gradient of a function:

$$U \equiv -G \cdot \sum_a \frac{M_a}{||\vec{x}_a - \vec{x}||} \Rightarrow \vec{g} = -\overrightarrow{\text{grad}}_x[U].$$ (1.1.2)

The particle P also exerts a gravitational force on the masses M_a. If the mass of P is small enough, this effect is negligible. In this case we call P a **test particle**.

From the work of appendix A, the motion of P can be deduced from the Lagrangian

$$L = \frac{1}{2}\left(\dot{x}^2 + \dot{y}^2 + \dot{z}^2\right) - U$$ (1.1.3)

where we have set $(\dot{\ }) = d(\ )/dt$.

We note that the Lagrangian is independent of the mass of P, in agreement with the Newtonian equivalence principle.

1.2 Motion of two bodies

1.2.1 Separation of the centre-of-mass motion

Consider two particles P_1 and P_2, of masses M_1 and M_2, which interact gravitationally. The Lagrangian can be written

$$L = \frac{1}{2}M_1\vec{V_1}^2 + \frac{1}{2}M_2\vec{V_2}^2 + G\,\frac{M_1 M_2}{r} \tag{1.2.1}$$

where $\vec{V_a}$ represents the particle velocities ($a = 1,2$) and r is the distance between P_1 and P_2.

If O is the origin of coordinates we set

$$\overrightarrow{OP}_1 = \vec{x}_1 \quad \text{and} \quad \overrightarrow{OP}_2 = \vec{x}_2$$

$$\vec{X} = \frac{M_1 \cdot \vec{x}_1 + M_2 \cdot \vec{x}_2}{M_1 + M_2}, \quad \vec{r} = \vec{x}_2 - \vec{x}_1. \tag{1.2.2}$$

The total mass is $M = M_1 + M_2$; we call $\mu = M_1 M_2 / M$ the **reduced mass**.

We express the Lagrangian (1.2.1) in terms of $\vec{X}$, $\dot{\vec{X}} = d\vec{X}/dt$, $\vec{r}$ and $\dot{\vec{r}} = d\vec{r}/dt$. The equations satisfied by $\vec{X}$ and $\vec{r}$ decouple:

$$M\frac{d^2\vec{X}}{dt^2} = \vec{0} \qquad \mu\frac{d^2\vec{r}}{dt^2} + G\frac{\mu M}{r^2}\frac{\vec{r}}{r} = \vec{0}.$$

The first equation shows that the centre of mass moves with constant velocity, while the second equation is that of a test particle of mass μ in the gravitational field of a central mass M. The latter equation describes the **relative motion**. For $M_1 \gg M_2$, the centre of mass of the system is almost the same as P_1; the relative motion is then the motion of P_2 in a Galilean frame centred on P_1.

The relative motion can be studied in spherical polar coordinates with the Lagrangian

$$L = \frac{1}{2}\left(\dot{r}^2 + r^2\dot{\theta}^2 + r^2\dot{\phi}^2\sin^2\theta\right) + \frac{GM}{r} \tag{1.2.3}$$

where we have set $\vec{r} = \{r\sin\theta\cos\phi, r\sin\theta\sin\phi, r\cos\theta\}$.

1.2.2 The relative motion

We find the equations satisfied by θ, ϕ and r.

θ equation

$$\frac{d}{dt}(r^2\dot{\theta}) - r^2\dot{\phi}^2\sin\theta\cos\theta = 0. \tag{1.2.4}$$

We choose the coordinate axes so that $\theta = \pi/2$ and $d\theta/dt = 0$ for $t = 0$. This equation and its derivatives then show that the time derivatives of θ to any order vanish for $t = 0$. As we can assume that the function $\theta(t)$ is analytic, we have

$$\theta(t) = \frac{\pi}{2}. \tag{1.2.5}$$

The relative motion therefore lies in a plane.

We could also assume that $r(t)$ is known and argue that (1.2.4) is then a second-order equation in θ. The initial values $\theta = \pi/2$ and $\dot{\theta} = 0$ give the unique solution (1.2.5).

ϕ equation

$$\frac{d}{dt}(r^2\sin^2\theta \, \dot{\phi}) = 0. \tag{1.2.6}$$

Using the relation $\theta = \pi/2$, we get

$$r^2\dot{\phi} = j \tag{1.2.7}$$

where j is a constant; μj is the angular momentum about the axis Oz.

r equation

The corresponding Lagrange equation can be replaced by the equation of conservation of total energy (cf. appendix A § A.4)

$$\frac{1}{2}(\dot{r}^2 + r^2\dot{\phi}^2) - \frac{GM}{r} = E \tag{1.2.8}$$

where E is a constant; μE is the total energy of relative motion, or the internal energy of the two-body system: kinetic + potential.

In this equation we replace $\dot{\phi}$ by j/r^2:

$$\frac{1}{2}\dot{r}^2 + \frac{1}{2}\frac{j^2}{r^2} - \frac{GM}{r} = E. \tag{1.2.9}$$

To determine the trajectory, we regard r as a function of ϕ and set $\sigma = 1/r$, $\sigma' = d\sigma/d\phi$:

$$\dot{r} = -\frac{\dot{\sigma}}{\sigma^2} = \frac{-\sigma'\dot{\phi}}{\sigma^2} = -j\sigma' \Rightarrow \frac{1}{2}j^2(\sigma'^2 + \sigma^2) - GM\sigma = E. \tag{1.2.10}$$

This equation becomes

$$d\phi = \frac{\pm d\sigma}{F(\sigma)} \quad \text{with} \quad F^2 = -\sigma^2 + \frac{2GM}{j^2}\sigma + \frac{2E}{j^2}. \tag{1.2.11}$$

The extreme values, $\sigma_\pm$, of σ are those for which

$$\frac{d\sigma}{d\phi} = 0 = F(\sigma_\pm), \quad \text{or} \quad \sigma_\pm = \frac{GM}{j^2}\left\{1 \pm \sqrt{1 + 2E\left(\frac{j}{GM}\right)^2}\right\}.$$

As F^2 is quadratic in σ we can write it as

$$F^2 = (\sigma_+ - \sigma) \cdot (\sigma - \sigma_-) \quad \text{with} \quad \sigma_- \leqslant \sigma \leqslant \sigma_+. \tag{1.2.12}$$

With a suitable choice of origin for the angle ϕ, (1.2.11) integrates to give

$$\begin{cases} \sigma = \dfrac{1}{r} = \dfrac{1}{2}(\sigma_+ + \sigma_-) + \dfrac{1}{2}(\sigma_+ - \sigma_-)\cos \phi \\[2mm] \dfrac{1}{2}(\sigma_+ + \sigma_-) = \dfrac{GM}{j^2} \\[2mm] \dfrac{1}{2}(\sigma_+ - \sigma_-) = \dfrac{GM}{j^2} \sqrt{1 + 2E \left(\dfrac{j}{GM} \right)^2}. \end{cases} \tag{1.2.13}$$

The trajectories are conics with one focus at the origin. $E > 0$ corresponds to a hyperbola, $E = 0$ to a parabola, and $E < 0$ to an ellipse.

1.2.3 Applications to photon orbits

We regard light as a jet of photons, massless particles of speed c. In Newtonian theory, the action of gravity on a test body gives an acceleration independent of the body considered. If this acceleration field applies also to massless particles it must influence the behaviour of light rays. We do not claim that this interpretation is contained in classical theory, but simply explore the possibility of such an extension, which picks out the acceleration field rather than the force field and the corpuscular aspect of light as opposed to its wave character.

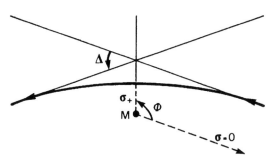

Fig. 1.2

Light deflection

The generalisation discussed above implies that light is deflected through an angle Δ when it passes close to a massive object such as the Sun (fig. 1.2).

We define the **Schwarzschild radius** of the Sun:

$$R_s = \frac{2GM}{c^2}. \tag{1.2.14}$$

M is the Sun's mass: $M \approx 2 \times 10^{30}$ kg, so that $R_s \approx 3 \times 10^3$ m.

The deflection $|\Delta|$ can be found by integrating (1.2.11):

$$\Phi = \int_0^{\sigma_+} \frac{d\sigma}{\sqrt{(\sigma_+ - \sigma)(\sigma - \sigma_-)}}$$

$$\cos \Phi = -\frac{(\sigma_+ + \sigma_-)}{(\sigma_+ - \sigma_-)} = -\sin \left(\frac{\Delta}{2}\right). \tag{1.2.15}$$

Using (1.2.13) we find Δ as a function of r_+. The deflection is maximal for $r_+ = R$, the Sun's radius. Under these conditions, with $R \approx 7 \times 10^8$ m, we have

$$|\Delta| \approx \frac{R_s}{r_+} = \frac{R_s}{R} \approx 0.88 \text{ seconds of arc}. \tag{1.2.16}$$

Consider a situation where the photon is very far from the central mass: with speed c the relation (1.2.8) gives $E = c^2/2$. Note that $E > 0$ implies $\sigma_- < 0$ and the corresponding angle ϕ becomes π, while σ_+ is positive at $\phi = 0$.
For $r = r_+$ the speed of a photon is $V = r_+ \dot{\phi}$, with $V^2 = c^2 (1 + R_s/r_+)$. The relation $r_+ > R$, with $R \approx 7 \times 10^8$ m $>> R_s$, gives $V \approx c$. Then (1.2.7) gives $j = r_+ c$. Using (1.2.13), we express σ_+ and σ_- as functions of r_+. The relation (1.2.15) with $r_+ = R$ gives the expression (1.2.16) for the deflection.

Light deflection by the Sun is actually observed; the measured value is twice the prediction here, but in agreement with general relativity.

Michell and Laplace's dark bodies

We consider a particle with radial velocity V in the spherically symmetrical gravitation field of a central mass. The particle will always recede if E is positive, or equivalently if its velocity exceeds the escape velocity $V_L = \sqrt{2GM/r}$, where r is its distance from the origin. Photons have speed c. They can leave the surface of an astronomical body and reach any point of space if $c > V_L$. For this, the body must not be too compact, i.e. its radius must exceed $R_s = 2GM/c^2$. If not, no photon will be able to reach infinity. There would then be regions of space in which the object would be invisible. General relativity implies a stronger result: a visible object must have a "radius" greater than its Schwarzschild radius R_s, and is then visible everywhere.

1.3 Local equations for the gravitational field

Newton's law of universal gravitation, together with the equivalence of gravitational and inertial mass, determines the acceleration field $\vec{g}$ once the matter distribution is known. Introducing the potential U and the mass density ρ of the matter distribution, Newton's law can be expressed as

$$\begin{cases} \vec{g} = -\overrightarrow{\text{grad}[U]} \Longleftrightarrow \overrightarrow{\text{curl}[\vec{g}]} = \vec{0} \\ -\text{div } [\vec{g}] \equiv \Delta U = 4\pi G \rho \end{cases} \tag{1.3.1}$$

Here $\vec{g}$, U and ρ are functions of position and possibly of time. The Laplace operator Δ is defined in the usual way: $\Delta(\ ) = \partial^2(\ )/\partial x^2 + \partial^2(\ )/\partial y^2 + \partial^2(\ )/\partial z^2$.

A point mass M gives a gravitational field $\vec{g}$ whose flux across an element of surface of area dS is

$$d\Phi = -\frac{GM}{r^2}\cos\theta\, dS = -GM d\Omega$$

where $d\Omega$ is the element of solid angle subtended by the surface element at the point mass (cf. fig. 1.3). Integrating, we find the total flux Φ of the gravitational field across a closed surface S as a function of the total mass M_{int} inside the volume V enclosed by S:

$$\Phi = -4\pi G M_{\text{int}} = -4\pi G \iiint_V \rho d^3 x.$$

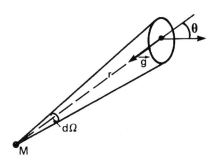

Fig. 1.3

Use of Gauss's theorem expresses Φ as a function of the integral of $\text{div}[\vec{g}\,]$ over V:

$$\Phi = \iiint_V \text{div}[\vec{g}\,]d^3 x = -4\pi G \iiint_V \rho d^3 x.$$

As this relation holds for any volume V, we get the local form (1.3.1) of the equations for the Newtonian gravitational field.

Conversely, for a spherically symmetric distribution of matter, possibly pointlike, we can show that (1.3.1) leads to the Newtonian expression (1.1.1) outside any domain enclosing the whole matter distribution. However, we have to demand that $\vec{g}$ vanishes at infinity. In cosmology, matter is not confined to a limited region of space. The boundary conditions above are then replaced by initial and symmetry conditions.

The equations (1.3.1) must be supplemented by boundary conditions and by requiring $\vec{g}$ to vanish at infinite distance from the matter distribution, which we assume localised in a region of finite size. We assume the gravitational potential U to vanish similarly.

The solution of Poisson's equation (1.3.1) is then

$$U(\vec{x}, t) = -G \iiint \frac{\rho(\vec{x}', t)}{\|\vec{x} - \vec{x}'\|}d^3 x'. \tag{1.3.2}$$

The integration is extended over all space, or more precisely over the bounded region in which ρ is non-zero: $d^3 x'$ is the volume element: $d^3 x' = dx'dy'dz'$.

The gravitational field at the instant t depends on the matter distribution at the same instant: the "propagation" of the gravitational field therefore occurs at infinite speed.

Chapter 2

Metric theories of gravity

A knowledge of special relativitistic kinematics and the summation convention for repeated indices is required for reading this chapter (I § 5.1.1).

The special theory of relativity abandons the Newtonian separation of space and time and introduces the concept of space–time. Events in space–time are specified by four coordinates $\{x^\mu\}$ with $\mu = 0, 1, 2, 3$. This theory postulates that the "ageing" of a clock is a quantity defined *a priori*, and measured by the proper time of the clock, which has a hereditary character. All these notions remain unchanged here. Special relativity also postulates the existence of Minkowski coordinates $\{x^{\mu'}\}$, in which the proper time $d\tau$ can be expressed in a simple form:

$$c^2 d\tau^2 = (dx^{0'})^2 - (dx^{1'})^2 - (dx^{2'})^2 - (dx^{3'})^2. \tag{2.0.1}$$

$d\tau$ is the proper time between two neighbouring events along the worldline of the clock considered, with coordinates $\{x^{\mu'}\}$ and $\{x^{\mu'} + dx^{\mu'}\}$. c is a fundamental velocity which experiment identifies with the speed of light in vacuo. This chapter does not postulate the existence of global Minkowski reference systems. The theories we consider thus constitute an extension of special relativity. This extension allows the introduction of the gravitational field.

The theory of electromagnetism was present at the birth of special relativity. By contrast, gravitation theory was absent, which led to a difficult period of ten years while general relativity was being developed.

The Coulomb interaction between two charges is carried by the mediation of the electromagnetic field. This satisfies propagation equations with finite speed, in the form of Maxwell's equations. It is at least "unappealing" to assume that gravity can propagate with infinite speed.

It is possible to generalise the equations for the gravitational field and cast them in a form analogous to Maxwell's equations. However, this generalisation is not satisfying. Such a theory would not explain the equality of inertial and gravitational masses. It would predict a precession of the motion of the planets, similar to the Sommerfeld precession in atomic physics. Precession is indeed observed, but the theoretical predictions do not account for more than a small fraction of the observed effect.

Clearly a new theory of gravity which would embody directly answers to these questions without the introduction of *ad hoc* assumptions would constitute progress.

The first satisfactory theory of this type was Einstein's general relativity, which was essentially completed in 1916.

Since 1960, several "competing" theories have been proposed. Experiments and observations have been devised to discriminate among them. Many theories have already been eliminated and today there is nothing that would cause one to doubt the validity of general relativity.

However, one should not imagine that all theoretical speculation has stopped. We only know that general relativity is accurate in the domain where it has been tested, and this only to the precision of the measurements. The theory also remains purely classical, although it is desirable that it should take quantum mechanics into account. This aim alone would modify the theory.

Among all current and future relativistic theories of gravity, a particularly important class is formed by **metric theories**, which we shall discuss in this chapter.

2.1 Einstein's equivalence principle

The Newtonian theory of gravity predicts, and experiment confirms, that the following proposition is true. Consider a small laboratory, in free fall, without communication with the outside world, such as a satellite. Within the laboratory, nothing will tell us that any of the bodies inside it is acted upon by a gravitational field, as they all feel the same gravitational acceleration. This is the Newtonian equivalence principle.

In such a satellite, the effect of the Earth's gravity may be ignored, and the laws of physics are those of special relativity. Equation (2.0.1) is therefore valid. We can therefore imagine a Minkowski reference frame equipped with clocks and measuring rods in free fall with the satellite; we refer to a Minkowskian **observer** in free fall. We could recognise such a reference frame by use of accelerometers and gyroscopes. However, the Minkowski coordinates $\{x^{\mu'}\}$ are not necessarily comoving with the satellite, which might rotate with respect to the reference frame. The Minkowski reference frame would give a description of the motion of objects in its neighbourhood even if they are not freely falling. To describe a small region of space–time "here" and "now" for example, we can introduce a freely falling Minkowskian observer who is "here" at the instant "now". Moreover, if motions with respect to this reference frame are slow, the observer can be identified with a Galilean observer in Newtonian theory.

Imagine an experiment carried out on Earth with ourselves as a terrestrial observer. The freely falling observer will regard us as accelerated upwards; he would regard what we call "the Earth's gravitational attraction" as in fact an inertial force, which from our viewpoint gives all bodies the same downwards acceleration. We should explain the same phenomenon in terms of the Newtonian law of universal gravitation in the Earth's field, Newton's second law of motion, and the equality of gravitational and inertial mass.

In reality, material particles fall towards the centre of the Earth. Their accelerations are not exactly parallel. If measurements are very accurate, it is possible to distinguish a gravitational field from an inertial field unless the laboratory is infinitesimally small. The Newtonian equivalence principle thus suggests that the physical laws of special relativity are valid in the presence of a gravitational field in an infinitesimally small laboratory.

This proposition constitutes **Einstein's equivalence principle**.

This principle can only hold for local laws. Thus we can deduce Maxwell's equations from this principle; but the field "clothing" a charged particle is not confined to an infinitesimal region, so we cannot assert that a charged particle will have the same free-fall motion as a neutral one.

The Einstein equivalence principle must be used with care. It does not allow us to construct the entire theory, and in this sense it plays the role rather of a guide. As an example, consider two point masses infinitesimally close in the Earth's gravitational field. In this case the direct application of Einstein's principle is impossible. The gravitational forces do not reduce to inertial ones, as the gravitational interaction between the two masses dominates as their separation decreases. The Einstein equivalence principle would be applicable if we could assume that objects could always feel gravitational fields without creating them. Such test bodies do not exist; but the concept allows us to describe many physical situations to a good approximation. We may then use the Einstein equivalence principle, and we shall do this frequently in the following.

The question arises of how much faith to place in a theory constructed on the basis of an academic principle. The answer lies in direct confrontation with experiment.

2.2 The idea of a metric

In developing the descriptive framework of the theory, we cannot limit ourselves to introducing local Minkowski reference frames; we must use coordinate systems in which physical laws can be formulated exactly, not merely in the limit in which the laboratory is infinitesimal. Let $\{x^\alpha\}$ be such a coordinate system. The proper time between two neighbouring events A and B with coordinates $x_A = \{x_A{}^\alpha\}$ and $\{x_A{}^\alpha + dx^\alpha\}$ on the worldline of a particle follows from equation (2.0.1) in the local Minkowski reference frame. In the coordinates $\{x^\alpha\}$, the line element has the form

$$c^2 d\tau^2 = g_{\alpha\beta}(x_A)dx^\alpha dx^\beta \tag{2.2.1}$$

where $g_{\alpha\beta}(x) = g_{\beta\alpha}(x)$ are the coefficients of the **metric** at the point with coordinates $x = \{x^\alpha\}$.

The ten functions $g_{\alpha\beta}(x)$ describe the inertial fields appearing in a reference frame comoving with the coordinate system we have introduced (cf. I § 7.3.3), as well as the gravitational field.

The absence of global Minkowski coordinate systems, such that $g_{\alpha\beta} = \eta_{\alpha\beta}$ everywhere, signals the presence of a gravitational field.

As in special relativity, we define $g^{\alpha\beta}(x)$ such that

$$g_{\alpha\mu}g^{\mu\beta} = \delta_\alpha{}^\beta = \delta^\beta{}_\alpha = g^{\beta\mu}g_{\mu\alpha} \text{ and } g^{\alpha\beta} = g^{\beta\alpha}. \tag{2.2.2}$$

Here too we use the rules for indices: summation of repeated indices, and the rule for raising and lowering indices (cf. I chapter 5).

Local Minkowski coordinates at the point A can be found from the $\{x^\alpha\}$ through the relations $x^{\alpha'} = F_A{}^{\alpha'}(x)$, where the four functions $F_A{}^{\alpha'}(x)$ depend on the event A considered.

We thus get

$$g_{\alpha\beta}(x_A) = g_{\beta\alpha}(x_A) = A^{\tau'}{}_\alpha(x_A)A^{\mu'}{}_\beta(x_A)\eta_{\tau'\mu'}$$

where $A^{\tau'}{}_\alpha(x_A) = (\partial F_A{}^{\tau'}/\partial x^\alpha)_A$ (cf. I chapter 7) and $\eta_{\tau'\mu'} = \text{diag}\,(1,-1,-1,-1)$.

The Einstein equivalence principle suggests that such formulae might hold at each point of space–time. However, there do not in general exist functions $x^{\tau'} = F^{\tau'}(x)$, independent of A, which allow us to write the coefficients $A^{\tau'}{}_\alpha(x)$ in the form $\partial x^{\tau'}/\partial x^\alpha$ for all x.

2.3 Free fall of a test particle

Consider the space–time with metric $g_{\alpha\beta}(x)$ in a global coordinate system $x = \{x^\alpha\}$. To find the motion of a test particle in free fall, we consider neighbouring events A and B on the particle's worldline. As A and B are arbitrarily close, we can appeal to the Einstein equivalence principle and regard the free fall motion as the motion of a free particle in special relativity. The motion is therefore such that the proper time between A and B is maximal. This condition can be written in an arbitrary coordinate system $\{x^\alpha\}$. The resulting equations are local; they do not depend on A and B. A and B may thus be arbitrarily close, i.e. chosen so that the Einstein equivalence principle holds, but the same equations would also be found by considering arbitrary events A and B.

We therefore consider two events A and B on the worldline of a test particle in free fall.

Let $x = \{x^\alpha\} = \{X^\alpha(u)\} = X(u)$ be the parametric equation of this worldline. The proper time τ_{AB} between the events A and B, calculated along this worldline, is

$$\tau_{AB} = \frac{1}{c} \cdot \int_{u(A)}^{u(B)} \sqrt{g_{\alpha\beta}[X(u)] \cdot \frac{d}{du}X^\alpha(u) \cdot \frac{d}{du}X^\beta(u)} \cdot du$$

where $u(A)$ and $u(B)$ are the values of the parameter u at A and B.

Requiring τ_{AB} to be extremal is equivalent to writing Lagrange's equations for the Lagrangian $\pounds$, with

$$\pounds = \sqrt{g_{\alpha\beta}[X(u)] \cdot \frac{d}{du}X^{\alpha}(u) \cdot \frac{d}{du}X^{\beta}(u)}. \tag{2.3.1}$$

The parameter u is arbitrary here; the results of appendix A § A.6 show that we may choose it to put the Lagrangian in the form

$$\blacklozenge \qquad\qquad L = g_{\alpha\beta}[X(u)] \cdot \frac{d}{du}X^{\alpha}(u) \cdot \frac{d}{du}X^{\beta}(u). \tag{2.3.2}$$

A parameter u of this type is called **affine**. The equations of motion follow by requiring

$$\mathscr{A} = \int_{u(A)}^{u(B)} L(u)du \tag{2.3.3}$$

to be extremal. We use the notation $(\ )_{,\alpha} = \partial(\ )/\partial x^{\alpha} = \partial_{\alpha}$. The corresponding Lagrange equations are

$$g_{\alpha\beta,\mu}\frac{dX^{\alpha}}{du}\frac{dX^{\beta}}{du} - 2\frac{d}{du}\left(g_{\alpha\mu}\frac{d}{du}X^{\alpha}\right) = 0 \tag{2.3.4}$$

which can be written

$$\blacklozenge \qquad\qquad \frac{d^2X^{\mu}}{du^2} + \{^{\mu}{}_{\alpha\beta}\}\frac{dX^{\alpha}}{du}\frac{dX^{\beta}}{du} = 0 \tag{2.3.5}$$

where $\{^{\mu}{}_{\alpha\beta}\}$ are the Christoffel symbols:

$$\blacklozenge \qquad\qquad \{^{\mu}{}_{\alpha\beta}\} \equiv \frac{1}{2}g^{\mu\tau}(g_{\tau\alpha,\beta} + g_{\tau\beta,\alpha} - g_{\alpha\beta,\tau}). \tag{2.3.6}$$

These curves, on which the proper time is extremal, are called **geodesics**.

We expand (2.3.4) using the following equations:

$$\frac{d}{du}\left(g_{\alpha\mu}\frac{dX^{\alpha}}{du}\right) = \frac{dg_{\alpha\mu}}{du}\frac{dX^{\alpha}}{du} + g_{\alpha\mu}\frac{d^2X^{\alpha}}{du^2};$$

$$\text{with } \frac{dg_{\alpha\mu}}{du} = g_{\alpha\mu,\beta}\frac{dX^{\beta}}{du}, \text{ we get}$$

$$\frac{1}{2}g_{\alpha\beta,\mu}\frac{dX^{\alpha}}{du}\frac{dX^{\beta}}{du} - g_{\alpha\mu,\beta}\frac{dX^{\beta}}{du}\frac{dX^{\alpha}}{du} - g_{\alpha\mu}\frac{d^2X^{\alpha}}{du^2} = 0,$$

$$\text{or } g_{\alpha\mu}\frac{d^2X^{\alpha}}{du^2} + \{_{\mu\alpha\beta}\}\frac{dX^{\alpha}}{du}\frac{dX^{\beta}}{du} = 0.$$

Contracting each side with $g^{\sigma\mu}$, we find (2.3.5).

We note that $L(u)du^2 = c^2d\tau^2$, where $d\tau^2$ is the square of the proper time between two neighbouring events with parameters u and $u + du$ on the worldline of the freely falling particle; L is thus positive or zero.

Replacing X^α by the solution of the equations of motion in the expression for L, we get a function of u; we can show that $L(u)$ is constant (cf. appendix A § A.4). We can distinguish two cases.

(1) For $L > 0$, we can introduce the parameter $s = \pm u\sqrt{L}$. The sign is chosen so that s increases as the particle ages. Then on the worldline of a freely falling particle we have $L(s) \equiv g_{\alpha\beta}(s) \cdot dX^\alpha/ds \cdot dX^\beta/ds = 1$, $ds = cd\tau$. The corresponding geodesic is called "timelike", as its tangent vectors $\{dX^\alpha/ds\}$ have positive pseudo-norm, like timelike vectors in special relativity.

(2) For $L = 0$, we see that $d\tau = 0$. This is a local property, and thus can be interpreted as in special relativity: the particle has zero rest-mass, and thus follows a "null geodesic".

One of the four equations (2.3.5) may therefore often be replaced by the first integral:

$$\blacklozenge \qquad \begin{cases} L = 0 \text{ for a zero rest-mass particle} \\ L = 1 \text{ for a massive particle.} \end{cases} \qquad (2.3.7)$$

The equations of motion do not require the introduction of a mass. The motions of test particles with the same initial conditions are identical whatever their masses. In Newtonian theory this property is a consequence of the equality of gravitational and inertial masses. Metric theories here imply an economy of concepts.

We can consider the case $L = -1$ from a purely geometrical point of view. The corresponding curves are called "spacelike geodesics"; they do not correspond to the space–time trajectory of any physical particle. The fundamental element ds^2 between two neighbouring points is negative. We define "the proper length" between these points: $d\ell = \sqrt{-ds^2}$.

2.4 Metric theories

2.4.1 Definition of metric theories

Consider two coordinate systems $\{x^\mu\}$ and $\{x^{\mu'}\}$. The proper time between two neighbouring events on the worldline of a particle is the result of the ideal measurement of a quantity which exists *a priori*; the proper time is thus numerically invariant under changes of coordinates:

$$g_{\alpha\beta}dx^\alpha dx^\beta = g_{\mu'\tau'}dx^{\mu'}dx^{\tau'} \qquad (2.4.1)$$

with $dx^\alpha = a^\alpha{}_{\mu'}dx^{\mu'}$, $a^\alpha{}_{\mu'} = \partial x^\alpha/\partial x^{\mu'}$.

The relation (2.4.1) satisfied by arbitrary values of $dx^{\mu'}$ implies that

$$g_{\mu'\tau'} = a^\alpha{}_{\mu'}a^\beta{}_{\tau'}g_{\alpha\beta}. \qquad (2.4.2)$$

The transformation laws for $g_{\alpha\beta}$ are analoguous to those found in chapter 7 of part I; by analogy we say that the $g_{\alpha\beta}$ represent the components of a tensor of rank 2: **the metric tensor**. Similarly we characterise tensors of arbitrary rank by the transformation laws of their components analogous to those established in chapter 7 of part I for arbitrary coordinate changes. In appendix B we make precise the nature of the vector spaces and tensor products involved.

In the present context, we can imagine a theory in which gravitation is described by the metric tensor and possibly other fields. There are two problems. These are:

(1) the nature of laws governing physical systems in a given gravitational field;

(2) the equations determining the gravitational field and the metric in particular, once the nature and distribution of the sources are known.

To treat the first problem, we consider theories satisfying the Einstein equivalence principle together with the body of interpretations which lead to the relations (2.2.1) and (2.3.5). In these theories, the effect of gravity is introduced solely through the functions $g_{\alpha\beta}(x)$ and their derivatives.

This property characterises metric theories.

> Metric theories are not the only possible ones; but they are a class whose validity is currently unquestioned. As examples of metric theories we may give Einstein's general relativity, which uses only the metric tensor, and the Brans–Dicke theory, which has an additional field and gives general relativity as a limiting case.

The second problem will be treated in the framework of general relativity, which is a metric theory using this tensor alone to describe the gravitational field. For both problems then, we assume in the following that the metric tensor gives a complete description of gravity.

2.4.2 Weak fields

If the gravitational field vanishes in a simply connected region of space–time, there exists a global Minkowski reference frame there. If the field is weak, without strictly vanishing, there exists a coordinate system covering this region such that

$$g_{\alpha\beta} = \eta_{\alpha\beta} + h_{\alpha\beta} \text{ with } |h_{\alpha\beta}| << 1 \text{ and } \eta_{\alpha\beta} = \text{diag}(1, -1, -1, -1). \quad (2.4.3)$$

Such a coordinate system is called **quasi-Minkowskian**. The existence of a quasi-Minkowskian coordinate system $\{x^{u}\} = \{x^0, x^k\}$ may be regarded as the definition of a **weak gravitational field** in a given region of space–time.

Let us assume that the gravitational field is weak. In any quasi-Minkowskian system, we can use Newtonian language and define the "speed of particles": $V^k = dx^k/dt$, with $ct = x^0$. Assume that all speeds are small compared with c. Under these conditions of weak field and small velocities, we may demand the validity of Newtonian theory provided that we interpret the coordinates as quasi-Galilean coordinates in Newtonian theory. This interpretation shows *a priori* why Newtonian theory is successful. Moreover it agrees with the Einstein equivalence principle in the sense that special relativity reduces to Newtonian theory in the limit of small velocities.

To illustrate this condition by an example, consider the Lagrangian (2.3.1). According to the results of appendix A the parametrisation of the solution curves of Lagrange's equations is arbitrary; here we set $u = t$. We neglect $|h_{\alpha\beta}|^2$, $|h_{\alpha\beta} \cdot V^k/c|$ and $|V^k V^j/c^2|^2$. We get

$$\pounds = c \left(1 + \frac{1}{2} h_{00} - \frac{1}{2} \frac{V^2}{c^2} \right) \quad \text{with} \quad V^2 = \sum_k V^k V^k. \qquad (2.4.4)$$

In the static case this Lagrangian gives the Newtonian equations of motion for

◆
$$h_{00} = \frac{2U}{c^2} \qquad (2.4.5)$$

where U is the Newtonian potential (1.3.2).

Thus the assumed interpretation fixes h_{00} to a first approximation.

As an exercise, show that the Lagrange equations deduced from (2.4.4) have acceleration $\{d^2 x^k/dt^2\} = \ddot{g} = -\overrightarrow{\text{grad}[U]}$.

Chapter 3

Tests of metric theories in the Solar System

Chapters 5, 6 and 7 of part I are not essential here: only equation (3.1.3) is needed.

This chapter has two aims. First we wish to show the body of interpretations of metric theories which gives Newtonian theory in the appropriate domain of validity, and second, we present several results offering experimental support for metric theories of gravity.

We shall confine ourselves here to test bodies, planets and photons, moving in the gravitational field of a central body such as the Sun or the Earth.

The mathematics is greatly simplified by the assumption of spherical symmetry. We shall adopt this assumption in calculating relativistic perturbations. To first approximation, these simply add to the Newtonian perturbations, which are often larger, caused by deviations from spherical symmetry such as the presence of other planets or solar oblateness.

3.1 Spherically symmetric, static space–times

Spherical symmetry is a spatial property, and in relativistic theories the notion of space has no absolute validity. We thus specify the set of events which can be interpreted as a space and endowed with the required property.

Consider coordinate systems $\{x^\alpha\} = \{ct, x, y, z\}$ such that the set, E_t, of events with the same coordinate t can be interpreted as a space, with the coordinate t interpreted as a time.

For a non-moving observer at $x^k = a^k$, with $k = 1$, 2 and 3, and a^k constant, only the coordinate t varies. The observer's proper time is $d\tau$ with $d\tau^2 = g_{00}(t, a^k)dt^2$. Thus g_{00} is necessarily positive. The coordinate t can be regarded as the time of a clock. This clock does not show its own proper time; its behaviour varies with t and also depends on its "position", a^k. This is the usual situation, as in reality we do not know how to construct ideal clocks which show their proper time correctly.

Now consider two neighbouring points of E_t, with coordinates $\{x^0, x^k\}$ and $\{x^0, x^k + dx^k\}$. As special relativity is locally valid, the line element between

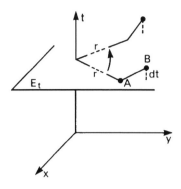

Fig. 3.1

these two events is negative if E_t can be interpreted as a space:

$$ds^2 = g_{kj}dx^k dx^j < 0.$$

In this case ds^2 is a symbol which does not represent the square of a physical quantity. The spatial distance between the two events is the proper distance dL, with $dL^2 = -g_{kj}dx^k dx^j = -ds^2$. This relation implies $g_{kk} < 0$, which in turn shows that x^k is a spatial coordinate.

> This definition is formal: nothing shows that the distance introduced in this way can be measured directly in the usual way. The space E_t is not in general the set of events that non-moving observers at $x^k = a^k$ would call "space". For this to be true, in accordance with special relativity, which is locally valid, we require that every interval $\{0, da^k\}$ should be orthogonal to the interval $\{dx^0, \vec{0}\}$. Under these conditions $g_{0k} = 0$. The distance dL may then be determined by measuring at $\{a^k\}$ the time $d\tau$ for photons to travel and return between non-moving objects at $\{a^k\}$ and $\{a^k + da^k\}$: $ds^2 = 0 \Rightarrow dL = cd\tau/2$.

The conditions under which t can be regarded as a time coordinate and x, y, z as space coordinates are thus

$$g_{00} > 0 \quad \text{and} \quad g_{kk} < 0 \text{ for all } k. \tag{3.1.1}$$

These conditions apply to the choice of coordinate systems and not to the space–time itself.

For the space–time to have spherical symmetry, there must exist among these coordinate systems one in which a spatial rotation, defined as an active transformation leaving invariant the quantity $(x)^2 + (y)^2 + (z)^2$, also leaves invariant the line element $ds^2 = g_{\alpha\beta}dx^\alpha dx^\beta$ between two arbitrary neighbouring events (fig. 3.1). We can show that there exist coordinates such that

$$\blacklozenge \qquad ds^2 = c^2 e^\nu dt^2 - e^\lambda dr^2 - r^2 d\theta^2 - r^2 \sin^2\theta \, d\phi^2 \tag{3.1.2}$$

where ν and λ are arbitrary functions of r and t.

In the diagram $\{x, y, z, t\}$, the geometry of the figure is completely determined up to a spatial rotation once we know

$$t, \; dt, \; r = \sqrt{x^2 + y^2 + z^2}, \; \vec{dx}^2 = dx^2 + dy^2 + dz^2$$
$$\text{and } \dot{x} \cdot \vec{dx} = xdx + ydy + zdz.$$

The line element ds^2 between A and B is assumed invariant under spatial rotations, so it must be expressible in terms of these quantities alone. Thus ds^2 has the form

$$ds^2 = F(r, t)c^2 dt^2 + G(r, t)dt(\dot{x} \cdot \vec{dx})$$
$$+ H(r, t)(\dot{x} \cdot \vec{dx})^2 + K(r, t) \cdot \vec{dx}^2.$$

We introduce "spherical" coordinates $x = r \sin\theta \cos\phi$, $y = r \sin\theta \sin\phi$, $z = r \cos\theta$:

$$ds^2 = Fc^2 dt^2 + rG dt dr + Hr^2 dr^2 + K(dr^2 + r^2 d\theta^2 + r^2\sin^2\theta \; d\phi^2).$$

In this expression, t is a time coordinate and r, θ and ϕ are spatial coordinates, which requires $F > 0$, $K < 0$ and $Hr^2 + K < 0$.

Making the change of variable $r' = r\sqrt{-K(r, t)}$, the line element takes the form

$$ds^2 = F'(r', t)c^2 dt^2 + G'(r', t)dt dr' - H'(r', t)dr'^2 - r'^2(d\theta^2 + \sin^2\theta \; d\phi^2)$$

with $F' > 0$ and $H' > 0$.

Another coordinate change $t = T(r', t)$ makes the term in $dt dr'$ vanish. We thus find

$$ds^2 = F''(r', t')c^2 dt'^2 - H''(r', t')dr'^2 - r'^2(d\theta^2 + \sin^2\theta \; d\phi^2)$$

which is the expression (3.1.2) up to notation.

For simplicity we make the additional assumption that the functions λ and ν are independent of t. The resulting static character of the gravitational field implies that experiments repeated with identical initial conditions give identical results whatever the epoch, measured by t, at which they are performed. This is a reasonable hypothesis as the duration of experiments in the Solar System is small compared with the characteristic evolution time of the Sun.

We may cast the metric (3.1.2) in so-called **isotropic** form:

$$ds^2 = B(r)c^2 dt^2 - A(r)(dr^2 + r^2 d\theta^2 + r^2\sin^2\theta \; d\phi^2) \tag{3.1.3}$$

with $A(r)$ and $B(r)$ positive.

This expression can be found from (3.1.2) up to notation by changing the radial coordinate to $r = r(r')$ with $e^{\lambda/2} dr/r = dr'/r'$.

3.2 Motion of a test particle

Given the line element (3.1.3), the motion of a test particle is found using (2.3.2) fom the Lagrangian

$$L = Bc^2 \left(\frac{dt}{du}\right)^2 - A\left\{\left(\frac{dr}{du}\right)^2 + r^2\left(\frac{d\theta}{du}\right)^2 + r^2\sin^2\theta\left(\frac{d\phi}{du}\right)^2\right\}. \qquad (3.2.1)$$

θ equation

$$\frac{d}{du}\left\{-2Ar^2\left(\frac{d\theta}{du}\right)\right\} + 2Ar^2\left(\frac{d\phi}{du}\right)^2\sin\theta\cos\theta = 0.$$

This equation can be analysed in the same way as the corresponding Newtonian equation (1.2.4). A suitable choice of coordinate system leads to the solution

$$\theta = \frac{\pi}{2}. \qquad (3.2.2)$$

In the following we replace θ by this value.

φ equation

This equation can be compared with the Newtonian equation (1.2.7), and leads to the first integral

$$A(r)r^2\frac{d\phi}{du} = J = \text{constant}. \qquad (3.2.3)$$

t equation

Integration of the t equation gives

$$c^2B(r)\frac{dt}{du} = K = \text{constant}. \qquad (3.2.4)$$

This equation has no Newtonian counterpart, as we used t as a parameter there.

r equation

From (2.3.7) this equation can be replaced by

$$c^2B(r)\left(\frac{dt}{du}\right)^2 - A(r)\left\{\left(\frac{dr}{du}\right)^2 + r^2\left(\frac{d\phi}{du}\right)^2\right\} = L \qquad (3.2.5)$$

with $L = 0$ for a massless particle, and $L = 1$ otherwise.

To facilitate comparison, we parametrise the worldline using t as in the Newtonian theory. We express $d(\ )/du$ as a function of $d(\ )/dt$ using (3.2.4):

$$
\left\{
\begin{aligned}
&\theta = \frac{\pi}{2} \\
&\left(\frac{A}{B}\right) r^2 \left(\frac{d\phi}{dt}\right) = j \\
&\left(\frac{A}{B^2}\right) \left\{ \left(\frac{dr}{dt}\right)^2 + r^2 \left(\frac{d\phi}{dt}\right)^2 \right\} - \frac{c^2}{B} = 2\mathcal{E}
\end{aligned}
\right.
\tag{3.2.6}
$$

where j and $\mathcal{E}$ are constants expressible as functions of L, K and J, with $\mathcal{E} = 0$ for a massless particle ($L = 0$).

We set $\sigma = 1/r$ and determine the equation obeyed by σ as a function of ϕ.

$$
\left\{ \frac{j^2}{2A} \right\} \left\{ (\sigma')^2 + \sigma^2 \right\} - \frac{c^2}{2B} = \mathcal{E}
\tag{3.2.7}
$$

with $\sigma' = d\sigma/d\phi$.

This equation may be compared with (1.2.10); it can be written as

$$
d\phi = \pm \frac{d\sigma}{F(\sigma)}
\tag{3.2.8}
$$

with

$$
F^2 = 2A\frac{\mathcal{E}}{j^2} + \frac{Ac^2}{j^2 B} - \sigma^2
$$

where A and B are regarded as functions of σ rather than r.

3.3 The Newtonian limit

3.3.1 Equations of motion and the Newtonian limit

In the Newtonian theory the spherical coordinates r, θ, ϕ in the Copernican reference frame centred on the Sun are determined along with the absolute time t by measurements and numerical calculations which ensure that these quantities have the usual physical interpretation. In relativity, the same measurements and calculations determine the coordinates of the events considered, without an *a priori* physical interpretation. In this coordinate system, regarded as Galilean, Newtonian theory is obeyed to a very good approximation by slow motions, such as those of the planets for example. We may therefore assume that the gravitational field is weak and identify the coordinate system with a quasi-Minkowskian system in spherical coordinates. To interpret the experimental results, we identify this coordinate system with the isotropic coordinate system giving expression (3.1.3).

In reality we must proceed differently. We have to postulate the validity of (3.1.3), study the motions of the planets theoretically, as well as the behaviour of the light rays which allow us to observe them, and the behaviour of the measuring apparatus. We could then determine the theoretical predictions from the measurements actually made and adjust the integration constants in the various equations so that the theory would agree as well as possible with experiment. Using such a method would lead to a "small" difference of interpretation between the coordinates determined with Newtonian hindsight and the coordinates leading to (3.1.3). These small differences would give small changes to the already small perturbations caused by metric theories compared with Newtonian theory. To a first approximation, such an approach is unnecessary.

We set

$$A = 1 + a(r) \quad \text{and} \quad B = 1 + b(r) \quad \text{with} \quad |a| << 1 \quad \text{and} \quad |b| << 1.$$

In the slow motion approximation, $(dr/dt)^2 << c^2$ and $r^2(d\phi/dt)^2 << c^2$, equations (3.2.6) give in first approximation

$$
\begin{cases}
\theta = \dfrac{\pi}{2} \\[2mm]
r^2 \dfrac{d\phi}{dt} = j \\[2mm]
\dfrac{1}{2}\left(\dot{r}^2 + r^2\dot{\phi}^2\right) + \dfrac{1}{2}\,c^2 b(r) = \dfrac{2\mathcal{E} + c^2}{2} \equiv E.
\end{cases}
\tag{3.3.1}
$$

These equations are identical with the Newtonian equations (cf. § 1.2), provided that

$$B \approx 1 - \frac{2GM}{rc^2} \tag{3.3.2}$$

where M is the mass of the central body creating the gravitational field: this is the Sun in the case of the planets. This result agrees with the expression (2.4.5).

In general relativity, the expression (3.3.2) relating g_{00} to the matter distribution can be found from the field equations.

3.3.2 The Einstein frequency shift

Study of the Newtonian limit leads to expression (3.3.2) characterising the gravitational field rather than the motions. We can use this form of $B(r)$ to study the ultrarelativistic motions performed by photons. The agreement between experiment and theory in the case of large velocities justifies the general framework we have introduced.

Consider a static light source S, with coordinates θ_s, ϕ_s, r_s.

This source emits light particles, or "photons", assumed to be massless particles, with a period T_s, measured at S (fig. 3.2).

A static observer O has coordinates $\theta_0 = \theta_s$, $\phi_0 = \phi_s$, $r_0 \neq r_s$.

Photons have $\mathcal{E} = 0$. From (3.2.6), those with radial trajectories move so that $j = 0$, with equation of motion

$$dt = \pm\sqrt{\frac{A}{B}} \cdot \frac{dr}{c}.$$

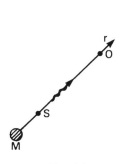

Fig. 3.2

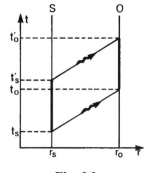

Fig. 3.3

Two successive photons are emitted at times t_s and t'_s towards the observer. They are detected at times t_0 and t'_0 (fig. 3.3). The equations of motion lead to the relation

$$\frac{T_s}{T_0} = \sqrt{\frac{B(r_s)}{B(r_0)}}. \tag{3.3.3}$$

On the source's worldline, with proper time $d\tau$, we have $ds^2 \equiv c^2 d\tau^2 = B(r_s)c^2 dt^2$. We deduce the period T_s:

$$T_s = \sqrt{B(r_s)}(t'_s - t_s).$$

In the same way we find

$$T_0 = \sqrt{B(r_0)}(t'_0 - t_0).$$

The equations of motion lead to the relation

$$t_0 - t_s = \int_{r_s}^{r_0} \sqrt{\frac{A}{B}} \frac{dr}{c} = t'_0 - t'_s.$$

We deduce $t'_0 - t_0 = t'_s - t_s$, which leads to (3.3.3).

A set of gas atoms, in free fall between collisions, emits spectral lines whose proper frequency as measured by a comoving observer we know from terrestrial measurements. The same atoms emit lines of the same proper frequencies on the Sun. This is an application of the equivalence principle in a laboratory of atomic dimensions over the very short time required for the emission of a wave train. To describe the mean effect we assume the emitting atoms and the observer are at rest in the isotropic coordinate system we have used. We thus know T_s. We measure T_0 by observing from Earth the line emitted at the Sun. We set $r_0 = \infty$ and $r_s = R$, the Sun's radius. Considering only the gravitational field of the Sun, we set

$$B(r_s) = 1 - \frac{2GM}{Rc^2} \qquad B(r_0) = 1$$

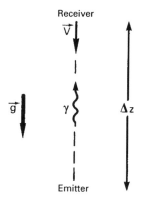

Fig. 3.4

where M is the mass of the Sun: $R \approx 7 \times 10^8$ m, $M \approx 2 \times 10^{30}$ kg. The relative frequency variation is found using (3.3.3).

$$\frac{\Delta v}{v} = \frac{T_0 - T_s}{T_s} \approx \frac{GM}{Rc^2} \approx 2 \times 10^{-6}.$$

This frequency shift constitutes the Einstein effect. Whatever the radiation considered the shift is towards longer wavelengths for $r_0 > r_s$. It is thus a "redshift"; in the contrary case we would have a blueshift. Although this terminology applies strictly to visible light, it is used for any wavelength.

The effect is difficult to detect for a source at the Sun (Brault, 1962), but has been convincingly demonstrated in the Earth's field (Pound and Rebka, 1960; Vessot and Levine, 1979). The discrepancy between the theoretical prediction (3.3.3) and the observed results is less than 2×10^{-4} in the most accurate experiments, and may be attributed to experimental error.

To explain the principle of the Pound–Rebka experiment we consider gamma-ray photons emitted by nuclear deexcitation, which move along a vertical in the Earth's gravitational field, on which an emitter and receiver are situated.

The emitter is a set of excited iron nuclei, static in the Earth's field, while the receiver is a set of iron nuclei in the ground state.

In this experiment (fig. 3.4), the spectral shift is to the red if the radiation rises. If the receiver is static the frequency of the emitted line is no longer equal to that of the absorption line. This effect can be compensated by the Doppler effect if the receiver moves downwards with speed V.

The radiation intensity is measured behind the receiver as V is varied. The value of V corresponding to the minimum intensity is precisely that which exactly compensates the gravitational redshift.

We can describe the experiment in the above notation by setting

$$B(r) = 1 - \frac{2GM}{rc^2} \text{ with } r = R + z.$$

Here, M is the mass of the Earth, R its radius and z the height of the point considered: $M \approx 6 \times 10^{24}$ kg, $R \approx 6.4 \times 10^6$ m. Expanding (3.3.3), we get

$$\frac{\Delta \nu}{\nu} \approx g\frac{\Delta z}{c^2} \approx \frac{V}{c} \text{ where } g = \frac{GM}{R^2} \approx 9.8 \text{ m s}^{-2}.$$

In the expression above, Δz is the height difference between the source and the observer, i.e. between the emitter and the receiver. Pound and Rebka's first experiment was carried out at Harvard, where the emitter and receiver were separated by a height difference $\Delta z \approx 22.5$ m. With this difference the relative frequency variation is of order 2×10^{-15}.

In the first experiment, the accuracy was no more than 1%.

To perform this experiment several conditions are required: the absorption and emission lines must be very narrow as we wish to measure a displacement from the line centre of order 2×10^{-15} (i.e. we want the receiver to absorb only resonant photons; this is required to determine the conditions of maximum absorption precisely). Iron nuclei have the required properties. They must form part of a crystal of sufficient mass that the recoil effect is negligible both in emission and in absorption (Mössbauer effect).

The measurements of Vessot and Levine used two very stable clocks, in the form of hydrogen masers, of which one stayed on the Earth while the other was launched in a rocket. The clocks measure the period of the same wave, emitted from the Earth. As the height difference is large, the effect is "noticeable".

3.4 The post-Newtonian approximation

3.4.1 Eddington–Robertson expansion

In the line element (3.1.3), $A(r)$ and $B(r)$ are dimensionless quantities which depend on the central mass M and possibly on the fundamental constants G and c. These quantities are close to unity because special relativity is well verified in the Solar System; they must be exactly equal to one if there is no gravitation, i.e. if $G = 0$, or $M = 0$ or $r = \infty$. The only dimensionless quantities that one can form with G, M, c and r are functions of GM/rc^2. We may thus assume that A and B can be expanded in powers of GM/rc^2.

We note that outside the Sun, $GM/rc^2 < 2 \times 10^{-6}$. Thus the first terms in these expansions must give any observable effects.

$$\begin{cases} A = 1 + 2\gamma\dfrac{GM}{rc^2} + 2\gamma' \left\{ \dfrac{GM}{rc^2} \right\}^2 + \dots \\[3mm] B = 1 - 2\alpha\dfrac{GM}{rc^2} + 2\beta \left\{ \dfrac{GM}{rc^2} \right\}^2 + \dots \end{cases} \tag{3.4.1}$$

We do not claim that all theories allow such expansions; we consider the set of theories for which this is true. With each of these theories is associated a set of parameters $\gamma, \gamma', \alpha, \beta$, with precise numerical values, depending on the theory considered. These are called parametrised post-Newtonian theories, or **PPN theories**.

The arguments above suggest that most reasonable theories are PPN theories. This is true of general relativity in particular.

The form (3.3.2) of $B(r)$ can be regarded as the starting-point for PPN expansions (3.4.1) with $\alpha = 1$. If M is the Newtonian mass of the central object, the value of α is determined by the hypothesis that Newton's theory is correct in the limit of weak fields and low velocities. For the Sun, the Newtonian value of M is determined by the motions of the planets; we do not claim that it is the sum of the masses of the constituent parts of the Sun, but simply that it is a coefficient characterising the Sun's gravitational field.

We will examine two experiments which measure γ and β. With these values known, PPN theories can then be compared with experiment.

The PPN formalism has been extended to cases where spherical symmetry does not necessarily hold; we shall not consider such cases here.

3.4.2 Trajectories in the Solar System

To determine the trajectories of test bodies in the Solar System we use the expression (3.2.8), in which we express A and B by means of the expansion (3.4.1) up to terms in $\{GM/rc^2\}^2$.

$$F^2(\sigma) = -\Omega^2\sigma^2 + \frac{2GM}{j^2}\left(1 + \frac{2E\gamma}{c^2}\right)\sigma + \frac{2E}{j^2} \tag{3.4.2}$$

where E and j are defined in (3.3.1) and

$$\Omega^2 = 1 - 2\left(\frac{GM}{cj}\right)^2\left\{2\frac{E\gamma'}{c^2} + 2 + 2\gamma - \beta\right\}. \tag{3.4.3}$$

$F^2(\sigma)$ is a quadratic in σ which is non-negative. We consider here a perturbation of Newtonian theory, i.e. we assume $\Omega \approx 1$. This implies $\sigma_- \leqslant \sigma \leqslant \sigma_+$ where we have defined $\sigma_\pm$ by the relations $F(\sigma_\pm) = 0$.

The relations (3.2.8) then imply that $d\sigma/d\phi = 0$ for $\sigma = \sigma_\pm$, so that the values $\sigma_\pm$ correspond to extrema of σ, just as in Newtonian theory (cf. § 1.2.2).

Equation (3.2.8) is then

$$\Omega d\phi = \pm\frac{d\sigma}{\sqrt{(\sigma_+ - \sigma)(\sigma - \sigma_-)}} \tag{3.4.4}$$

σ can thus be expressed in terms of $\Omega\phi$ using the first of the Newtonian relations (1.2.13).

$$\sigma = \frac{1}{2}(\sigma_+ + \sigma_-) + \frac{1}{2}(\sigma_+ - \sigma_-)\cos\Omega\phi. \tag{3.4.5}$$

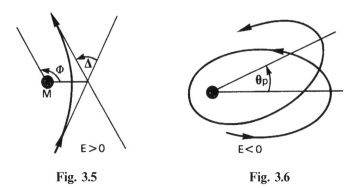

Fig. 3.5   Fig. 3.6

We distinguish two general cases:
(1) $E > 0$
In this case $\sigma_+\sigma_- = -2E/(\Omega j)^2$ is negative; so $|\sigma_+ + \sigma_-| < |\sigma_+ - \sigma_-|$. The quantity $\sigma = 1/r$ vanishes for values $\phi = \Phi$ corresponding to the asymptotes (fig. 3.5):

$$\cos \Omega\Phi = -\frac{(\sigma_+ + \sigma_-)}{(\sigma_+ - \sigma_-)}.$$ (3.4.6)

(2) $E < 0$
In this case σ does not vanish. The motion is that of an ellipse precessing through an angle θ_p at each revolution (fig. 3.6):

$$\Omega(2\pi + \theta_p) = 2\pi$$ (3.4.7)

where $(2\pi + \theta_p)$ is the angle through which ϕ must increase for σ to take the same value.

3.4.3 Deflection of light rays by the Sun

Consider the trajectory of a photon. This corresponds to $\mathscr{E} = 0$ or $E = 1/2 \cdot c^2$. Keeping only the lowest-order correction in G in the expression for $F(\sigma)$, we find

$$F^2(\sigma) = -\sigma^2 + \frac{2GM}{j^2}(1+\gamma)\sigma + \frac{c^2}{j^2}.$$ (3.4.8)

This expression is identical to the Newtonian expression (1.2.11) provided we replace G by $G(1 + \gamma)$. The deflection of light rays by the Sun is thus

$$|\Delta| = \frac{2GM}{r_+c^2}(1+\gamma).$$ (3.4.9)

We have already stated that observational results give a value twice that calculated in section 1.2.3. The measured value of γ is thus

$$\gamma = 1.$$ (3.4.10)

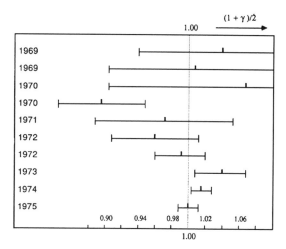

Determination of $(1 + \gamma)/2$ by measurement of the deflection of radio waves by the Sun, with both the mean values and "error bars". The main uncertainty comes from the solar corona.

Fig. 3.7

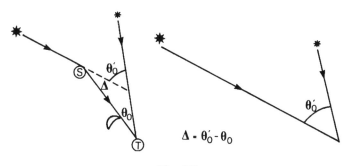

Fig. 3.8

The value $\gamma = 1$ is that predicted by general relativity. The accuracy to which γ is known is of order 10^{-3}.

The first observations of the deflection of light rays by the Sun were made in the optical in 1919 (expedition to Sobral in Brazil). The accuracy was barely 20%. The observations were performed during a solar eclipse, in which the Moon blocked the Sun's light. Under these conditions it is possible to photograph the star field near the Sun. Comparison with plates of the same region when the Sun is absent gives a measure of the deflection (fig. 3.8).

Since these first results the development of radio astronomy has greatly increased the accuracy and removed the necessity to observe during an eclipse. Moreover the coefficient γ can be measured in another way, through the **Shapiro effect** which is seen in a delay of radar echoes.

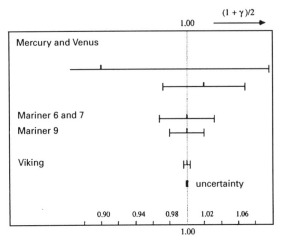

Determination of $(1 + \gamma)/2$ through the Shapiro effect before the 1980s. Radar echoes may be produced by the planets themselves (Mercury and Mars), by space probes equipped with transponders (*Mariner 6* and 7), or by transponder satellites around Mars or deposited on it (*Mariner 9* and *Viking*).

Fig. 3.9

When radio waves travel across the Solar System from the Earth to Mars, and are then reflected from Mars back to Earth, the travel time measured by the terrestrial observer depends on γ. It is larger than the time predicted by Newtonian theory. The extra delay is smaller than about 10^{-4} s. It has nevertheless been measured to high precision, particularly during the Viking mission.

The deflection of light rays has several observable consequences, such as gravitational lenses.

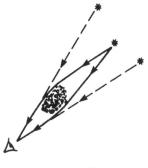

Fig. 3.10

A gravitational field created by a central mass acts like a light deflector. A galaxy or cluster of galaxies can thus deviate the light rays from a source situated behind them. The terrestrial observer can see this source in differing directions (fig. 3.10). This phenomenon, first demonstrated in 1980, is known as **gravitational lensing**.

3.4.4 Perihelion advance of the planets

The motion of the planets is decribed by Newton's theory to very high precision. According to (3.3.1), mE represents the Newtonian mechanical energy of a planet of mass m in the Sun's gravitational field. For the non–relativistic case of interest to us here, $mE \ll mc^2$, or $E/c^2 \ll 1$. Thus for sufficiently slow motion, expression (3.4.3) for Ω^2 reduces to

$$\Omega^2 = 1 - 2\left(\frac{GM}{cj}\right)^2 (2 + 2\gamma - \beta)$$

leading to the value of θ_p from (3.4.7).

$$\begin{cases} \theta_p = \left(\frac{2\pi GM}{Lc^2}\right)(2 + 2\gamma - \beta) \\ \frac{1}{L} = \frac{GM}{j^2} \approx \frac{1}{2}(\sigma_+ + \sigma_-) = \frac{1}{2}\left(\frac{1}{r_+} + \frac{1}{r_-}\right). \end{cases} \qquad (3.4.11)$$

Mercury makes about 415 revolutions per century on a quasi-elliptical trajectory with semi-major-axis $a = (r_+ + r_-)/2 \approx 5.8 \times 10^{10}$ m and eccentricity $e = |(r_- - r_+)/(r_- + r_+)| \approx 0.21$. The advance of the perihelion of Mercury was observed and largely explained in the 19th century by Newtonian effects. However, Le Verrier pointed out the presence of a discrepancy, today estimated at about 43 seconds of arc per century, in the middle of the 19th century. If we attribute this extra advance to the precession effect studied here we can find β by using $\gamma = 1$. With the data given, we can verify that $\beta \approx 1$. Various measurements lead to the value $\beta = 1$ with an accuracy currently of order 5×10^{-3}. General relativity predicts $\beta = 1$.

The perihelion advance of the planets (cf. 3.4.11) was calculated assuming that the Sun's gravitational field is static and spherically symmetric. This is not true to the highest precision: the Sun rotates slowly on its axis (with a velocity of order 1.9 km s^{-1} at the equator) and is flattened along this axis (the relative difference between the apparent diameter along the axis and that along the equator is of order 10^{-5}). Thus to a second approximation the Sun is merely axisymmetric.

The solar oblateness and the resulting deviation from spherical symmetry produce a Newtonian precession of the motions of those planets whose trajectories lie close to the Sun's equatorial plane. The precession is equal to that given by general relativity, θ_p, multiplied by a dimensionless quantity λ_p:

$$\lambda_p = \frac{2 + 2\gamma - \beta}{3} + J_2 \frac{R^2}{R_s L}$$

where R and R_s are the radius and Schwarzschild radius of the Sun respectively, and J_2 is a dimensionless number characterising the quadrupole moment of the Sun: J_2 would vanish if the Sun had complete spherical symmetry.

A gravitational lens produces several images (cf. § 3.4.3). If the arrangement of observer, deflector and object is axially symmetric, the image is a circle. Under certain conditions close to axial symmetry the circle becomes an arc. In the general case we observe several images usually with differing intensities.

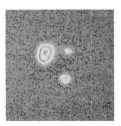

Fig. 3.11

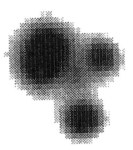

Fig. 3.12
(© CNRS/CFHT)

Fig. 3.13

The image 3.11 was taken in 1985 by C. Vanderriest and G. Wlerick at the CFHT, the Canada–France–Hawaii telescope. We see three images of the quasar P.G. 1115+080 ("quasar" stands for "quasi-stellar", denoting extremely distant intrinsically bright objects related to active galaxies). The image is in false colour, each shade corresponding to a photon density. Here there are isophotes separating the two regions.

The images 3.12 and 3.13 are more recent. They were taken at the CFHT by C. Vanderriest (Observatoire de Paris/CNRS) in March 1992. On 3.12 we can just see the lensing galaxy, while the brighter image appears as itself possibly formed from two very close images.

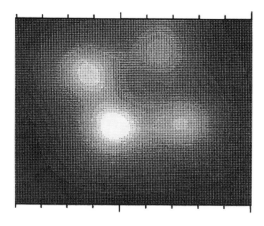

Fig. 3.14

This picture was taken at the CFHT (cf. fig. 3.15) and analysed in 1989 by M.C. Angonin and C. Vanderriest (Observatoire de Paris and CNRS), M. Remy and J. Surdej (Institut d'Astrophysique de l'Université de Liège). These are four images of the quasar H 1413+117, more commonly called the "four-leafed clover".

The optical spectrum of a source is represented by plotting the flux per wavelength interval $(Wm^{-2}\text{Å}^{-1})$ against wavelength.

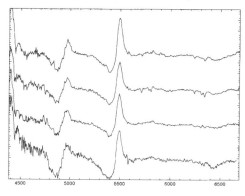

Fig. 3.15

The curves above show the spectra of each of the four components of the four-leafed clover between 4500 Å and 6500 Å. For clarity the spectra have been displaced vertically by varying amounts. We note the extreme similarity of the four spectra. This similarity shows that these are four images of the same source and that we are dealing with a gravitational lens (cf. § 3.4.3). Some small differences are present, which can be interpreted physically. The reader should consider the problem of comparing spectra when they are of sources with very different intensities.

The precise measurement of the precession angles of several planets would determine both J_2 and $2 + 2\gamma - \beta$. However, in the Solar System, only Mercury has a sufficiently eccentric orbit and short enough period for measurements of θ_p to reach the required accuracy.

General relativity is characterised by the values $\beta = \gamma = 1$. Within experimental error, Mercury's perihelion advance is explicable in general relativity with $J_2 = 0$. Under these conditions we can only regard this measurement as a positive test of general relativity if J_2 is not too large.

At the end of the 1960s, some observations suggested that J_2 might be as large as 3×10^{-5}. At that time, more than a century after Le Verrier, measurement accuracy was of order 1%. Such a value of J_2 would imply $(2 + 2\gamma - \beta)/3 > 1$: thus general relativity appeared to fail and controversy arose. A little later, further observations led to an estimate for J_2 of order 10^{-6}; the controversy died down. Since then many observations in very diverse domains have "confirmed" general relativity. Rather than measuring J_2 to test the theory, nowadays the perihelion advance of Mercury is measured and general relativity used to work out the size of J_2.

Using the results above, determine the maximum value of J_2. For the Sun, assume $R \approx 700\,000$ km, $R_S \approx 3$ km and for Mercury $L \approx 55 \times 10^6$ km with $\lambda_P = 1 \pm 10^{-2}$.

Chapter 4

Newtonian cosmology

This chapter does not use any relativistic concepts. However, the reader should be familiar with the Newtonian theory of gravity introduced in the first chapter of part two.

Cosmology is "the study of the Universe in its entirety". One would hope that this study would provide a framework largely determining physical laws and elucidating their true nature. In fact the current approach consists mainly of using observations and established laws to describe the behaviour of the Universe on the largest scale.

Cosmology is now studied using relativistic theories. However, several concepts such as the redshift or big bang are already present in the Newtonian theory which is the subject of this chapter.

4.1 Distance measurements

4.1.1 Units and orders of magnitude

The Earth's orbit about the Sun is almost circular, with a radius 1.5×10^{11} m. This distance is called an **astronomical unit**.

If one moves away from the ecliptic plane along the axis of the Earth's orbit, the apparent diameter of this orbit decreases. At a distance of about 3.1×10^{16} m, an astronomical unit subtends one second of arc. This distance is called a **parsec** (pc). Kiloparsecs and megaparsecs are also used, with 1 kpc $= 10^3$ pc and 1 Mpc $= 10^6$ pc.

In one year, light travels about 9.5×10^{15} m; this distance is called a **light year** (lyr).

The nearest stars are at several light years. The Sun is a star whose nearest neighbour α Centauri is at just over 1 pc (about 4 lyr).

Stars are distributed in galaxies. Our galaxy is the **Milky Way**. This is flattened like a pancake, with a central bulge. The maximal thickness is 3 kpc and the diameter about 30 kpc, while the mass is $1.5 \times 10^{11} M_0$ ($M_0 =$ solar mass $\approx 2 \times 10^{30}$ kg).

In our immediate vicinity, at a distance of barely 1 Mpc, the Andromeda galaxy, the Magellanic Clouds and a few small galaxies form the **Local Group**.

Fig. 4.1

This galaxy, NGC 2903, is in the constellation of Leo at 5 Mpc from
our Galaxy, the Milky Way. Several spiral arms emerge from the bright
nucleus. Two symmetrical exterior arms are very extended but scarcely
visible, one particularly so.

(© CNRS/OHP)

Groups of galaxies are called **clusters of galaxies** when they are very rich.
The Virgo cluster contains several thousand galaxies. It is at the centre of the
local supercluster, which comprises several tens of clusters.

The Local Group is at the edge of the Virgo supercluster whose centre is
at about 15 Mpc.

On the largest scale, of order 100 to 1000 Mpc, the Universe is usually
regarded as homogeneous and isotropic.

The schematic description we have just given should not obscure the great
diversity of astronomical objects. The masses of galaxies vary widely; some
of them are one thousand times smaller than our own, while others are
much larger. They have various morphologies; elliptical galaxies or spiral
galaxies like our own, with spiral arms. The stars populating the galaxies
also divide into very different families. The Sun has a mean density of order
1.4×10^3 kg m^{-3}, while neutron stars are close to the nuclear density of order
5×10^{17} kg m^{-3}. Moreover, stars, galaxies, groups and clusters are not the
only astrophysical objects.

Although it is well established that galaxies have a tendency to cluster as
we have indicated, this is not a rule without exceptions. Many galaxies are
isolated. Further, study of the matter distribution in the observable Universe
has shown the existence of "voids", clusters of clusters and walls and filaments.

Fig. 4.2

This star cluster, called NGC 6205 or M 13, is 8 kpc from us. Its diameter is 11 pc. The number of stars is estimated at 200 000 and there are about 50 stars per cubic parsec.
Currently 120 globular clusters of this type are known within a sphere of 30 kpc diameter about our Galaxy.

(© CNRS/OHP)

At the cosmological scale which concerns us here we assume that matter is distributed as a homogeneous fluid: **the cosmological fluid.**

4.1.2 The distance scale

The distance, d, of a nearby object can be found by trigonometric methods.

Consider a triangle whose points are the object under study and the positions of the Earth at two epochs separated by six months. The base of the triangle is the diameter of the Earth's orbit, which is known. Measuring the base angles, simple trigonometry gives the distance d of the object. If d becomes too large, $d > 30$ pc, this method becomes very inaccurate.

Within the region accessible to the method just described, it is possible to study many stars and classify them according to their "spectral type". Stars like the Sun which are the site of thermonuclear reactions fusing hydrogen to make helium constitute the **main sequence**. If we know the distance, d, of a star we can find its luminosity L, expressed in watts, as a function of its flux ℓ measured at Earth, expressed in W m^{-2}:

$$L = 4\pi d^2 \ell. \tag{4.1.1}$$

Independently, E. Hertzsprung and H. N. Russell showed at the beginning of the century that the luminosity L of main sequence stars depends on their spectral type. Once this dependence is empirically established, we can then determine L by observing the spectral type, and then measure ℓ at the Earth and use the relation (4.1.1) to find d. This method gives the distances of nearby galaxies, as we can regard all the stars as having the same distance.

Most main sequence stars are not very luminous: this method can only be used for distances smaller than 10^5 pc.

In this region of the Universe we observe variable stars called *Cepheids* and *RR Lyrae* stars. These stars, when studied in galaxies at known distances, have a luminosity L which we can find by measuring the flux ℓ at Earth. This luminosity varies with a period T of a few days or tens of days; L and ℓ are thus periodic functions of time. There is in fact a precise relation between $L(t)$, $\ell(t)$ and the period T of $\ell(t)$. This allows us to find $L(t)$ by observing $\ell(t)$. Using (4.1.1) we find the distance of the variable star. This method allows one to reach distances of order 4 Mpc.

Many galaxies are observed in the regions accessible to these methods. It is noticeable that the brightest stars have practically the same absolute luminosity L in all these galaxies. This result is not very surprising if we assume similar mechanisms for star formation. One can then determine the distances of far-off galaxies by observing their brightest stars. As their absolute luminosities are known one can measure their flux at Earth and again use (4.1.1). This method allows one to reach distances of order 30 Mpc.

At very large distances it is impossible to resolve stars in galaxies, and one can only resolve galaxies in clusters. The distance to a cluster may then be found by using the fact that the brightest galaxies in clusters have similar luminosities in all clusters. This value is found by observations of nearby clusters whose distances are known by one of the preceding methods. In this way one can reach distances of order 10^4 Mpc.

Distance measurements are subject to many errors. In using (4.1.1) we are making two assumptions: isotropy of the radiation and the absence of absorption of light. Isotropy is reasonable for the objects considered; but by contrast the absorption of light by the interstellar medium is not negligible. In directions within the plane of the Galaxy absorption may be so great that observation of distant objects is impossible. Estimates of the absorption vary and remain uncertain. Another uncertainty comes from statistical effects. It is for example possible that the luminosity of galaxies does not have an identical maximum for all clusters but has a distribution such that large values have low probability without actually being excluded. In this case the probability of a high value increases with the richness of the cluster and the number of clusters considered. This mechanism results in a underestimate of the luminosity of the brightest galaxies in distant clusters and thus an underestimate of their distances. We have discussed several methods and several reference objects which allow us to construct a distance scale. These objects, called **standard candles**, and these methods, are not unique. The results obtained in various ways are not always in good agreement. A further difficulty appears if we study objects which are not themselves standard candles. If for example a galaxy is seen "in" a cluster this does not necessarily mean that it is a cluster member; it may simply lie on the same line of sight, in front of or behind the cluster. Also, if a quasar is seen close to a galaxy this does not mean that it is really close. To prove an association of the quasar and the galaxy would require one to observe several cases and calculate the probability of such observations assuming that galaxies and quasars are independently distributed over the sky. If the probability is low we may be justified in concluding that the observed associations are not simply random alignments but result from real proximity in space.

4.2 The Hubble effect

Every atom or molecule is characterised by its emission spectrum: the intensity of the emitted light as a function of the frequency or wavelength in vacuum. The emission results from deexcitation of atoms previously excited by collisions caused by thermal agitation for example. The set of lines characterising an element constitute its **spectrum**. When we observe the light from a distant object, we recognise the spectra of various elements. The wavelengths are, however, modified. This modification may result from the Doppler effect caused by the motion of the source relative to the observer (cf. I chapter 3). Taking the mean of observations of sources at the same distance d reveals a *systematic* spectral shift towards longer wavelengths, i.e. a *redshift*. This effect was discovered by Hubble in 1929 and results from expansion.

We set $Z = (T_0 - T_s)/T_s = (\lambda_0 - \lambda_s)/\lambda_s$, where T_0 and T_s are respectively the observed and emitted periods and λ_0 and λ_s the corresponding vacuum wavelengths ($\lambda = cT$). T_0 is directly observed, while for T_s we assume the value measured on the Earth from a source of the same type. Thus for a given astronomical source Z is a known quantity.

It is found that Z increases linearly with the distance d of the source: $Z = \alpha d$. This is the **Hubble effect**. By *interpreting* this effect as a Doppler effect we find the recession velocity V of the particles of the cosmological fluid.

Assuming $V/c << 1$, we find $Z \approx V/c$ and

$$V = H_0 d \qquad (4.2.1)$$

where H_0 is the **Hubble constant**.

Light is emitted at time t_s and observed at time t_0. In (4.2.1), the distance d is determined using (4.1.1); it is $c(t_0 - t_s)$. For nearby objects, $Z << 1$, d is practically equal to the distance at the time t_0; similarly, V is almost the recession velocity at that time.

Estimates of H_0 vary from 40 km s^{-1}/Mpc to 100 km s^{-1}/Mpc. For numerical estimates we shall adopt the order of magnitude

$$H_0 = 75 \text{ km s}^{-1}/\text{Mpc} = 2.4 \times 10^{-18} \text{ s}^{-1}. \qquad (4.2.2)$$

Hubble's law, $Z = \alpha d$, gives the distance of an object whose Z value we have measured. However, we must measure the systematic effect and separate it from the effect connected with the proper motion of the object under study. This is possible when the systematic effect is dominant (very distant objects). When this is not true, in estimating the value of Z for a cluster for example, one must use means over several objects of the cluster.

The distance indicator given by Hubble's law is not always in obvious accord with other estimates. Some quasars have large redshifts, suggesting large distances. Observing their fluxes at Earth, one can deduce their luminosities L; these are comparable to those of galaxies. These objects are variable. Collective mechanisms cannot work within the quasar with speeds greater than c. If T is the timescale for significant changes in intensity, the object cannot be larger than cT. Quasars thus appear to be much smaller than galaxies. Simple-minded reasoning might suggest that the quasar luminosity ought to be much smaller than that of a galaxy, and therefore that it should be much nearer. This disagreement revolves around the origin of quasar luminosities, although a minority of astronomers have questioned the cosmological origin of such large redshifts.

4.3 Reinterpretation of the gravitational field equations

To describe the large-scale behaviour of the Universe and Hubble's law, we assume that gravitational interactions alone are involved.

In a Galilean reference frame, $\{\Omega; \vec{i}, \vec{j}, \vec{k}\}$, a gravitational field reduces to an acceleration field, independent of the test particle considered, because of the equivalence of gravitational and inertial mass.

Consider an observer O, comoving with the cosmological fluid, who refers space to the frame $\{O; \vec{i}, \vec{j}, \vec{k}\}$. As he is carried along by the cosmological fluid, this observer is not necessarily Galilean. The acceleration of a test particle with respect to the observer's frame is

$$\vec{\gamma} = \vec{g} - \frac{d^2 \overrightarrow{\Omega O}}{dt^2}.$$

We have already given the local formulation of the laws of gravity (§ 1.3). Noting that $d^2\overrightarrow{\Omega O}/dt^2$ depends only on time we easily verify that the equations obeyed by $\vec{\gamma}$ are the same as those obeyed by $\vec{g}$. The only change comes from the differing behaviour of $\vec{g}$ and $\vec{\gamma}$ at infinity, far from any matter. In this case $\vec{\gamma}$ does not tend to zero. The acceleration field $\vec{\gamma}$ is the only observable quantity for the observer O. We can call it the "gravitational field with respect to O" and use the notation $\vec{g}$ now for this field in which we include the inertial acceleration $-d^2\overrightarrow{\Omega O}/dt^2$. From (1.3.1), the equations are

♦ $$\text{div} [\vec{g}] = -4\pi G\rho, \quad \overrightarrow{\text{curl} [\vec{g}]} = \vec{0} \tag{4.3.1}$$

where ρ is the mass density and G the Newtonian constant of gravitation.

We postulate the validity of equations (4.3.1) in the cosmological frame; however, we abandon the condition that $\vec{g}$ should vanish at infinity for the reasons explained above, and because matter is distributed throughout the Universe.

4.4 Mass conservation and equations of motion

In the frame associated with the observer O the density is a function of position and time. We introduce the velocity field with respect to this frame, $\vec{V} = \vec{V}(x, y, z, t)$: $\vec{V}$ is the velocity of a particle of the cosmological fluid at the point $\{x, y, z\}$ at time t.

The equation of mass conservation is

♦ $$\frac{\partial \rho}{\partial t} + \text{div} [\rho \vec{V}] = 0 \tag{4.4.1}$$

with $\text{div} [\vec{A}] \equiv \partial_k A^k$.

To write Newton's second law we must follow the motion of a particle. This particle passes from the position $\vec{r} = \{x, y, z\}$ to the position $\vec{r} + \vec{dr} = \vec{r} + \vec{V}dt$ in the interval dt. Its velocity changes from the value $\vec{V}[\vec{r}, t]$ to the value $\vec{V}[\vec{r} + \vec{dr}, t + dt]$. Its acceleration is then $(\vec{V}[\vec{r} + \vec{dr}, t + dt] - \vec{V}[\vec{r}, t])/dt$. Newton's second law is then given by equating the acceleration found in this way and that resulting from gravity:

♦ $$\frac{\partial \vec{V}}{\partial t} + (\vec{V} \cdot \vec{\nabla}) \cdot \vec{V} = \vec{g} \tag{4.4.2}$$

with $\vec{V} \cdot \vec{\nabla} \equiv V^k \partial_k$.

Equations (4.3.1), (4.4.1) and (4.4.2) are the fundamental equations of Newtonian cosmology.

4.5 Model of the expanding Universe

There is no reason to assume that we are privileged observers. The Hubble effect must therefore be general, and all particles of the cosmological fluid recede from each other. A simple way of taking this into account is to assume the relation

$$\overrightarrow{OM} = S(t) \cdot \overrightarrow{OM_0} \quad \text{with} \quad S(0) = 1 \tag{4.5.1}$$

where M is a particle of the fluid, and $\overrightarrow{OM_0}$ is the vector $\overrightarrow{OM}$ at time $t = 0$, conventionally chosen as the present epoch. Thus $\overrightarrow{OM_0}$ is a constant vector and $S(t)$ a function of time called the **scale factor**, such that $S(0) = 1$ by convention.

The expression (4.5.1) is universal as it is satisfied whatever the fluid particles P and Q considered:

$$\overrightarrow{PQ} = \overrightarrow{OQ} - \overrightarrow{OP} = S(t) \cdot \overrightarrow{OQ_0} - S(t) \cdot \overrightarrow{OP_0} = S(t) \cdot \overrightarrow{P_0 Q_0}.$$

The observer O is therefore not picked out by (4.5.1).

The velocity field can be found by differentiating (4.5.1):

$$\overrightarrow{V} = H \cdot \overrightarrow{OM} \quad \text{with} \quad H(t) = \frac{dS}{dt} \cdot \frac{1}{S}. \tag{4.5.2}$$

This equation explains the Hubble effect if this is interpreted as a Doppler effect. $H(0) = H_0$ is then the Hubble constant at the present epoch. As the spectral shifts are to the red, H_0 is positive, implying $(dS/dt)_0 > 0$.

The law (4.5.1) cannot hold unless equations (4.3.1), (4.4.1) and (4.4.2) hold also.

Assuming homogeneity at each instant, $\rho = \rho(t)$, we find div $[\rho \overrightarrow{V}] = 3H\rho$. From equation (4.4.1) we get $(1/\rho)d\rho/dt + 3H = 0$ with $H = (1/S)dS/dt$, or

♦
$$\rho = \frac{1}{S^3} \rho_0 \tag{4.5.3}$$

where ρ_0 is a constant giving the mass density of the Universe at the present epoch.

Writing the equations of motion (4.4.2):

$$\frac{1}{S} \cdot \frac{d^2 S}{dt^2} \cdot \overrightarrow{OM} = \dot{\overrightarrow{g}}. \tag{4.5.4}$$

The field equations (4.3.1) reduce to the single equation

$$\frac{3}{S} \cdot \frac{d^2 S}{dt^2} = -\frac{4\pi G \rho_0}{S^3}. \tag{4.5.5}$$

This equation has the first integral

♦
$$\frac{1}{2} \left(\frac{dS}{dt} \right)^2 - \frac{4\pi G \rho_0}{3} \frac{1}{S} = K \tag{4.5.6}$$

where K is a constant of integration expressible in terms of current quantities:

$$K = \frac{(H_0)^2}{2} - \frac{4\pi G \rho_0}{3}. \tag{4.5.7}$$

Thus the proposed model is compatible with the fundamental equations if

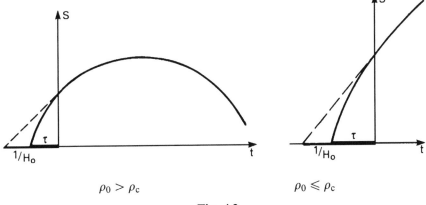

$$\rho_0 > \rho_c \qquad\qquad\qquad \rho_0 \leqslant \rho_c$$

Fig. 4.3

S satisfies equation (4.5.6), which is known as **Friedmann's equation**; ρ and $\vec{g}$ can then be found using (4.5.3) and (4.5.4).

Equation (4.5.6) is well known in another context. Consider a projectile with purely radial velocity, acted upon solely by the gravity of a spherically symmetric central mass M. Conservation of mechanical energy gives the relation

$$\frac{1}{2}\left(\frac{dr}{dt}\right)^2 - \frac{GM}{r} = E = \text{const.}$$

where r is the distance from the centre. This is the same equation as (4.5.6) up to notation. For $E \geqslant 0$, r increases indefinitely, while for $E < 0$ the projectile falls back after a certain time.

We can read off the shape of the function $S(t)$ from this discussion. In fig. 4.3 we have used the fact that at the current epoch $t = 0$ the observed spectral shifts are to the red, so that $dS/dt > 0$.

4.6 Discussion of the Newtonian expansion model

The above model gives a critical density ρ_c:

$$\rho_c = \frac{3(H_0)^2}{8\pi G} \approx 10^{-26} \text{ kg m}^{-3}. \tag{4.6.1}$$

For $\rho_0 \leqslant \rho_c$, the current expansion will continue for ever, but slow down. For $\rho_0 > \rho_c$ the expansion will slow down enough to be stopped; there will then be a contraction phase.

Estimates of ρ_0 from current observations give a value below $0.1\rho_c$. The problem is not settled, as estimates of ρ_0 are currently contradictory. Many indices suggest that the mass of the observable Universe cannot be more than a few per cent of the total mass: under these conditions ρ_0 might exceed ρ_c. This is the problem of the "missing mass", or "dark matter".

It is usual to introduce $q_0 \equiv -\{S \cdot (d^2S/dt^2)/(dS/dt)^2\}_{t=0}$, where q_0 is the **deceleration parameter**. Using (4.5.5) at the current epoch as well as (4.6.1), we find

$$\rho_0 = 2q_0\rho_c. \tag{4.6.2}$$

Determinations of q_0 are currently too uncertain to allow a clear decision as to whether the expansion will continue for ever or not.

4.7 Evolution of the Universe

Whatever the future of the Universe, it is clear that the model studied here requires the existence in the past, $t < 0$, of an epoch when $S(t)$ was very small. At that epoch ρ was very large (cf. 4.5.3). This was true of the entire Universe, and not some part of it which could be regarded as the centre.

To derive equation (4.5.4) we assumed that the cosmological fluid was *incoherent* and affected only by gravity; this assumption must break down once the matter density becomes too high and the pressure is no longer negligible.

The evolution of the Universe, and in particular its first moments, are described in **the standard model**. In this scenario, the Universe in the past had an epoch of very high density and temperature, the latter exceeding 10^{12} K. Matter did not appear as now, but rather as matter waves. After this initial epoch, which is still difficult to describe, the expansion began. This is known as the *big bang*.

The big bang occurred everywhere simultaneously in the Universe. The expansion was accompanied by a rapid drop in temperature; particles similar to those produced in accelerators appeared. The Universe continued to expand, the temperature, and thus the interaction energies, decreased producing a more stable situation. After about three minutes helium nuclei formed, explaining the abundance of this element, which could not have all been produced by thermonuclear reactions in the centres of stars. Much later, 10^5 to 10^6 years after the initial explosion, the temperature fell to about 4000 K, and hydrogen atoms formed, and from being an opaque plasma, the Universe became transparent.

Until that point electromagnetic radiation had been in thermal equilibrium with matter. After this point it was absorbed very little, and the thermal histories of matter and radiation followed separate paths. Today we still observe the radiation emitted in these last moments.

This emission was produced everywhere 10^{10} years ago. Thus it reaches us from every direction, from a distance of order 10^{10} light years. The Hubble effect shifts all the frequencies to the red, lowering the apparent radiation temperature to 2.7 K without changing the thermal nature of the

A point of the celestial sphere, i.e. of the sky, is specified by its latitude and longitude (called "declination" and "right ascension"). A few meridians and parallels are shown on the figure below. Our Galaxy is situated along the equator, its centre coinciding with the centre of the map.

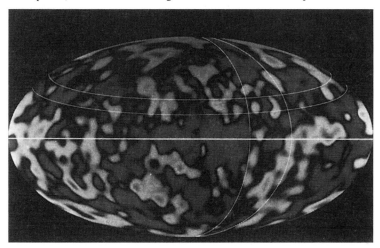

Fig. 4.4

The temperature of the relict radiation was measured in every direction (cf. § 4.7); it is shown on the original picture as colours, here represented as dark and light zones, which stand for cool and hot regions. The mean temperature is 2.73 K, and the relative fluctuations around this mean are less than 10^{-4}. Most of the patchiness represents experimental noise. However, for the first time, numerical analysis of the results appears to show that the background radiation itself has a certain irregularity whose amplitude is compatible with the fluctuations necessary for galaxy formation. Future observation should confirm these results and provide a map of the fluctuations.

The data shown on this picture from 1992 were collected by the DMR (Differential Microwave Radiometer) of COBE (COsmic Background Explorer), a NASA satellite launched in 1989 and operated by the Goddard Space Flight Center. They were analysed by the DMR team under G. F. Smoot.

(Photograph: NASA/Goddard Space Flight Center)

spectrum. Observations show that this radiation is very isotropic. However, the anisotropies required for the matter condensations which gave rise to the galaxies were observed in 1992 (fig. 4.4).

After this epoch of decoupling of matter and radiation, condensations formed and gradually produced the present contents of the Universe.

The great successes of the standard model were to predict the existence of the relict radiation and to explain the abundance of light elements. However, it is very important to make direct observations of various ill-understood mechanisms in the distant past, such as galaxy formation for example. We can do this by observing very distant objects, as their light has a finite velocity.

It is possible that in the near future, space-borne telescopes will allow such observations.

One of the most important questions is to estimate the age of the Universe τ; i.e. the time between the big bang and the present epoch. In fig. 4.1 we can see that $\tau < 1/H_0$; numerical solution of the equations of the current model shows that $1/H_0$ is an order of magnitude estimate for τ if q_0 has values compatible with observation.

The complicated processes of the first moments last for times much shorter than $1/H_0$ and thus do not affect the estimate of τ. For $H_0 \approx 2.4 \times 10^{-18}$ s^{-1} we find that the age of the Universe is of order 1.3×10^{10} years. This is of the same order as the age of the oldest galaxies, estimated by several methods. This is a satisfying result in view of the extreme simplicity of the model considered here.

Chapter 5

Relativistic cosmology

In the last chapter we introduced the concept of a cosmological fluid describing the "mean" behaviour of the Universe on a large scale. For observers moving with this fluid, the Universe has the same appearance in every direction. This property, extended to all possible observations, constitutes the **cosmological principle**. This is the starting point for the models we shall study in this chapter. First we must give a relativistic formulation of the cosmological principle. We shall establish the form of line element compatible with this principle, before interpreting the observations.

5.1 Discussion of the cosmological principle and the line element

We assume the existence of a cosmological fluid describing the Universe on the largest scale. The objects we observe do not precisely follow streamlines of this fluid as they have proper motions, like smoke particles in an air current. To discover properties of the cosmological fluid we must take the mean of many observations.

The distribution of matter in the Universe is discontinuous on the small scale and continuous on the large scale. Similarly a gas appears as a continuous fluid on the macroscopic scale and discontinuous on the molecular scale. However, we know how to connect the two descriptions. The situation is different in relativistic cosmology: the hypothesis of a cosmological fluid is an approximation which we do not yet know how to check.

To interpret the cosmological principle, we assume the existence of a coordinate system $\{x, y, z, t\}$ comoving with the fluid, such that E_t, the set of events with the same time coordinate t, is a homogeneous and isotropic "space".

These simplifying hypotheses lead to the following expression for the line element in the comoving coordinate system:

$$ds^2 = c^2 d\tau^2 = c^2 dt^2 - R^2(t)A^2(r)(dx^2 + dy^2 + dz^2) \tag{5.1.1}$$

where τ is the proper time and $r = \sqrt{x^2 + y^2 + z^2}$.

The functions $R(t)$ and $A(r)$ are chosen positive by convention. We set $S(t) = R(t)/R(t_0)$ where t_0 is interpreted as the present epoch. The function $S(t)$ is arbitrary for the moment, and is called the **scale factor**.

The coordinate system is comoving with the cosmological fluid: $dx = dy = dz = 0$ for a particle of this fluid. The time coordinate t thus represents the proper time of a fluid particle. It is called the **cosmological time**.

The line element has explicit spherical symmetry about the origin. According to the cosmological principle the observer at the origin with worldline $x = y = z = 0$ is not privileged. Every observer who is comoving with the fluid can be taken as at the origin of a suitably chosen coordinate system in which the line element has the form (5.1.1). This property assures the homogeneity of space and isotropy with respect to each of its points; it leads to an expression for $A(r)$ compatible with the model, namely

$$\blacklozenge \qquad\qquad A(r) = \frac{1}{1 + \dfrac{k}{4}r^2} \qquad\qquad (5.1.2)$$

with only the three possibilities $k = 1$, -1 or 0. The corresponding metric is called "Robertson–Walker" or "Friedmann–Robertson–Walker" (FRW). In this chapter we shall not use the explicit form of $A(r)$. We shall prove (5.1.2) in the following chapter.

To justify the form (5.1.1) of the line element we proceed by stages.

(1) We consider any observer moving with the cosmological fluid. His spatial coordinates x^k therefore remain constant. We choose two neighbouring events A and B on this observer's worldline, whose time coordinates are separated by $dz^0 = cdt$. We consider a neighbouring event D of A, belonging to the same space E_t (cf. fig. 5.1a). The difference of spatial coordinates between A and D is dy^k. As special relativity is locally valid, we can consider a local Minkowski frame associated with the observer at the point A. E_t can be interpreted as the corresponding space iff $(\overrightarrow{AB}, \overrightarrow{AD}) = 0$ for all D; i.e. $g_{\alpha\beta}dz^\alpha dy^\beta = 0$ for all dy^β and dz^α satisfying $dy^0 = 0$ and $dz^k = 0$. We thus get the relation $g_{0k} = 0$, so that

$$ds^2 = g_{00}c^2 dt^2 + g_{jk}dx^j dx^k.$$

(2) For each observer moving with the cosmological fluid the space E_t has the same local properties: mass density and temperature depending only on t. These quantities can be measured, so two different observers can decide if two events on their worldline belong to the same space E_t. As all observers are equivalent, their proper time between E_t and E_{t+dt} is the same. This implies that g_{00} is independent of the spatial coordinates of the observer. Thus we can define a new chronology $T = T(t)$ such that $dT^2 = g_{00}dt^2$. The spaces E_t can be labelled by T and written E_T. The line element then becomes

$$ds^2 = c^2 dT^2 + g_{jk}dx^j dx^k$$

where g_{jk} is now regarded as a function of T and x^k.

(3) We consider two fluid particles M_1 and M_2 in E_T with spatial coordinates differing by dx^k.

The spatial distance of the two particles L is such that

$$L^2 = -g_{ik}dx^i dx^k.$$

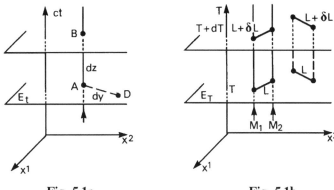

Fig. 5.1a   **Fig. 5.1b**

The quantities x^k and dx^k do not vary with T as the coordinate system is comoving with the cosmological fluid. However, between T and $T + dT$, the distance L varies by δL (cf. fig. 5.1b):

$$\delta(L^2) = 2L\delta L = -\frac{\partial g_{ik}}{\partial T}\, dT\, dx^i dx^k.$$

The isotropy and homogeneity of space imply that $\delta(L)$ has the same value whatever the particles considered, i.e. whatever the values of x^k and dx^k, if the initial distance L is the same: $\delta L = f(L, T)dT$. If L is infinitesimal, expansion to first order gives $\delta L = \{f(0, T) + f'(0, T)L\}dT$, with $f(0, T) = 0$; we deduce that $\delta(L^2)/L^2$ has the form $\delta(L^2)/L^2 = 2\delta L/L = F(T)dT$, implying the relation $\partial g_{jk}/\partial T = F(T)g_{jk}$; integrating, we get $g_{jk} = R^2(T)\Gamma_{jk}$ where Γ_{jk} is independent of T. The choice of a positive function $R^2(T)$ is purely conventional; in this case the sign of g_{jk} is that of Γ_{jk}. The line element can thus be written

$$ds^2 = c^2 dT^2 + R^2(T)\Gamma_{jk}dx^j dx^k$$

(4) The worldline of the origin, $x^k = 0$, is that of a particular observer comoving with the cosmological fluid. The isotropy of E_T for this observer implies that there exists a spatial coordinate system $X^k = F^k(x^j)$ such that the line element is invariant under rotations, transformations preserving the value of $r^2 = (X^1)^2 + (X^2)^2 + (X^3)^2$ in E_T. In spherical coordinates the line element can thus be written $ds^2 = c^2 dT^2 + R^2(T)\{G(r)dr^2 + K(r)r^2 d\theta^2 + K(r)r^2\sin^2\theta d\phi^2\}$ (cf. II § 3.1).

(5) Finally, we can make a last change of variables, $r = r(\rho)$, so that $\rho^2 G(r)\left(\dfrac{dr}{d\rho}\right)^2 = K(r)r^2$, and set $A^2(\rho) = -G[r(\rho)]\left(\dfrac{dr}{d\rho}\right)^2$. Then we have

$$ds^2 = c^2 dT^2 - R^2(T)A^2(\rho)\{d\rho^2 + \rho^2\sin^2\theta d\phi^2 + \rho^2 d\theta^2\}.$$

Using a "Cartesian" coordinate system we find (5.1.1) up to notation. We note that the coordinate changes made in (2), (4), and (5) above preserve the comoving character of the coordinates.

The line element (5.1.1) gives the equations of free-fall motion in the comoving coordinates. We verify that the particles of the cosmological fluid fall freely: $x^k = $ constant is a solution of (2.3.5). This shows that on the present model only gravity governs the motion of the cosmological fluid. This does not exclude a possible pressure, which will affect the functions $R(t)$ and $A(r)$.

5.2 Redshifts

Consider the metric (5.1.1) in spherical coordinates

♦ $$ds^2 = c^2 dt^2 - R^2(t)A^2(r)\left\{dr^2 + r^2\sin^2\theta d\phi^2 + r^2 d\theta^2\right\}. \qquad (5.2.1)$$

We choose the origin of coordinates to coincide with ourselves the observer. Placing the observer at the origin does not give him any privileged position as the space is homogeneous; we have fixed the particular coordinate system.

We consider a distant source with constant coordinates θ_s, ϕ_s, r_s, emitting periodic light signals received by the observer.

As an exercise one can verify that the motion of these photons is characterised by the relations

$$\theta = \theta_s, \quad \phi = \phi_s, \quad c^2 dt^2 = R^2(t)A^2(r)dr^2.$$

To establish this result, use the equations of motion (2.3.5) and (2.3.7) found in the second chapter of part II.

Two successive signals are emitted at times t_s and t_s', and received at times t_0 and t_0' (fig. 5.2).

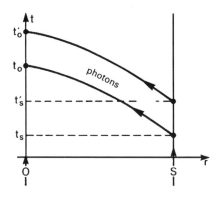

Fig. 5.2

The equations of motion give

$$\int_{t_s}^{t_0} c\,\frac{dt}{R(t)} = -\int_{r_s}^{0} A(r)dr = \int_{t_s'}^{t_0'} c\,\frac{dt}{R(t)} \qquad (5.2.2)$$

which leads to

$$\int_{t_s}^{t_s'} c\,\frac{dt}{R(t)} = \int_{t_0}^{t_0'} c\,\frac{dt}{R(t)}. \qquad (5.2.3)$$

The period of the signal measured by the observer is $T_0 = t'_0 - t_0$, while the period of the emitted signal measured at the source is $T_s = t'_s - t_s$. The function $R(t)$ varies very little over the period of the light signal and may be regarded as constant in the integrals (5.2.3) above. We thus get

$$\blacklozenge \qquad\qquad \frac{T_0}{T_s} = \frac{R(t_0)}{R(t_s)} \equiv 1 + Z. \qquad\qquad (5.2.4)$$

Thus the redshift, $T_0 > T_s$, is explained by the relation $R(t_0) > R(t_s)$.

5.3 Luminosity and redshift

5.3.1 Lengths, areas and volumes

In the laboratory we measure areas without indulging in cosmological considerations. Here we wish to relate the measured quantities to the choice of coordinates so as to interpret observations.˜

Consider the neighbourhood of an event O with coordinates $\{t_0, x_0, y_0, z_0\}$. To a first approximation $R(t)$ and $A(r)$ can be regarded as constants A_0 and R_0 in this neighbourhood.

We set $X = R_0 A_0(x - x_0)$, $Y = R_0 A_0(y - y_0)$, $Z = R_0 A_0(z - z_0)$. The line element is then to a first approximation Minkowskian: $ds^2 \approx c^2 dt^2 - (dX^2 + dY^2 + dZ^2)$. The coordinates $\{X, Y, Z\}$ are thus interpreted in the limit of small velocities as the Newtonian spatial coordinates of a Galilean system. The "distance" between two neighbouring points of the space E_t, separated by dx, dy, dz, is $dD = \sqrt{dX^2 + dY^2 + dZ^2} = R_0 A_0 \sqrt{dx^2 + dy^2 + dz^2}$.

To calculate a path length in E_t, we place a set of observers along it who each measure the element of length in their neighbourhood by the usual methods. We then add all these elementary results. In the limit where the observers are infinitesimally close to each other, we integrate dD over the path considered. The length found in this way is called the "proper length" of the path.

> We assume here that the "usual methods" give the value dD. This holds if we measure lengths by timing the round trip (out and back) of a light signal between two points. By contrast the use of measuring rods introduces the problem of the behaviour of the rods when they are transported over distance to be compared.

Using this method we can define in E_t the minimum distance of a point M from the origin O. This distance D is called the "proper distance", and depends on the coordinate r_0 of the point M:

$$D = R(t) \int_0^{r_0} A(r)dr = D(r_0).$$

In reality D is the proper distance from M to the origin; we use the shorter expression "proper distance" as we regard ourselves as an observer situated at the origin.

As an exercise, prove the above relation.

The ratio of the proper distances at times t_1 and t_2 of a particle of the cosmological fluid is thus $D_1/D_2 = R(t_1)/R(t_2) = S(t_1)/S(t_2)$. Comparing this relation and that deduced from (4.5.1), we see that the scale factors of the Newtonian and relativistic models play the same role.

Remark: The minimum distance between two points in E_t is an intrinsic quantity which does not depend on the coordinate system used. Thus, if we take the origin at M, the point O will have coordinates r'_0. As the metric retains its form, the function $D : r' \mapsto D(r')$ remains unchanged, as does its numerical value: $D(r_0) = D(r'_0)$. Thus $r_0 = r'_0$.

To determine the areas of a surface in E_t we operate in the same way: knowing this surface in the coordinate system x, y, z, we find the element of area as in Euclidean space, and the result is then multiplied by $R^2(t)A^2(r)$. Integration over the surface considered then gives the desired result.

As an example, consider the sphere of given "radius" r. On this sphere the element of area is $R^2(t)A^2(r)r^2\sin\theta\,d\theta d\phi$.

The area of the sphere is then

$$S = \int_0^\pi d\theta \cdot \int_0^{2\pi} R^2(t)A^2(r)r^2\sin\theta\,d\phi = R^2(t)A^2(r)4\pi r^2. \qquad (5.3.1)$$

Volumes are found similarly. The volume element is first calculated in the coordinate system x, y, z which is assumed Euclidean. It is then multiplied by $R^3(t)A^3(r)$. A volume integration gives the result. For example, for the sphere of radius r_0, we get

$$V = \iiint_0^{r_0} R^3(t)A^3(r)r^2\sin\theta\,d\theta d\phi dr = 4\pi R^3(t)\int_0^{r_0} A^3(r)r^2 dr.$$

Here too the various expressions are the sums of elementary measurements made locally by many infinitesimally close observers, by the usual methods.

5.3.2 Luminosity distance

In the Newtonian cosmology of the preceding chapter we introduced the relation (4.1.1) expressing the luminosity L of an object as a function of the flux ℓ at Earth and its distance d; thus $L = 4\pi d^2\ell$. The use of this relation allows one to estimate distances by measuring ℓ, if we can estimate L independently. In relativistic models we may use this relation to define a distance. This is called the **luminosity distance**, and we write it as d_L. The question now arises of expressing d_L as a function of r.

We take the origin at the light source. The photons which it emits at time t_s have equations of motion $r = f(t)$ which are the same whatever their direction. They therefore cross the sphere of radius r_0 simultaneously (fig. 5.3).

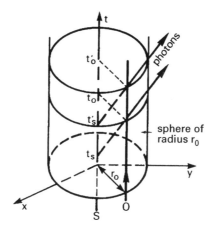

Fig. 5.3

Assume that N photons were emitted between t_s and t'_s. They reach the sphere between t_0 and t'_0. A calculation identical to that of the section above (§ 5.2) leads to the relation $t'_0 - t_0 = (t'_s - t_s)(1 + Z)$. For static observers at $r = r_0$ and $r = 0$ the time intervals corresponding to the passage and emission of the N photons are precisely $t'_0 - t_0$ and $t'_s - t_s$. Moreover, the frequency of each photon is redshifted, so the energy of the photons for an observer on the sphere is thus the energy measured at emission divided by $1 + Z$. Hence the luminosity P at the sphere is

$$P = \frac{L}{(1+Z)^2} = L \left(\frac{R(t_s)}{R(t_0)} \right)^2 \tag{5.3.2}$$

where L is the luminosity measured near the source. The flux ℓ measured at the sphere is P/S, where S is the area of the sphere.

Using expression (5.3.1) for S we find

$$L = (1+Z)^2 S\ell = \ell 4\pi A^2(r)r^2 \frac{R^4(t_0)}{R^2(t_s)}. \tag{5.3.3}$$

In this equation r is the radial coordinate r_0 of the observer. We can make a spatial coordinate change such that the line element keeps the form (5.1.1) and the new coordinates are centred at the observer. This possibility is a consequence of the spatial homogeneity of the model. Under these conditions r represents the radial coordinate r_s of the source (cf. remark p. 144). With the observer at the origin we thus get

$$d_L = r_s A(r_s) \frac{R^2(t_0)}{R(t_s)}. \tag{5.3.4}$$

The luminosity distance and proper distance are not equal. This inequality reflects the fact that the notion of distance is not intrinsic in relativity, and that a precise definition is needed whenever we introduce this concept.

5.3.3 Luminosity–redshift relation

For studies of our neighbourhood we assume that the observer is at the origin, and we expand $R(t)$ in a Taylor series:

$$R(t) = R(t_0) \left\{ 1 + H_0(t - t_0) - \frac{1}{2} q_0 H_0^2(t - t_0)^2 \right\} \qquad (5.3.5)$$

with
$$H_0 \equiv \left(\frac{1}{R} \frac{dR}{dt} \right)_{t=t_0} \quad \text{and} \quad q_0 \equiv -\frac{1}{H_0^2} \left(\frac{1}{R} \frac{d^2 R}{dt^2} \right)_{t=t_0}.$$

To lowest order in $t_0 - t_s$ the relations (5.2.4) and (5.3.3) give

$$Z = H_0(t_0 - t_s) \quad \text{and} \quad L = \ell 4\pi A^2(r_s) r_s^2 R^2(t_0)(1 + 2Z).$$

Under the same conditions the relation (5.2.2) becomes

$$c \frac{(t_0 - t_s)}{R(t_0)} \approx r_s A(0) \approx r_s A(r_s).$$

These equations immediately give

$$\ell = L \frac{H_0^2}{4\pi c^2 Z^2}. \qquad (5.3.6)$$

One can verify that this relation is analogous to the Newtonian relation deduced from the model of the last chapter. For this reason H_0 is called the "Hubble constant at the present epoch" in both models, even though the interpretation of the redshift is quite different in each case. Expansion to second order introduces the "deceleration parameter" q_0, obtained from $S(t)$ in the same way as the deceleration parameter in the Newtonian theory.

$$\ell = LH_0^2 \frac{1 + (q_0 - 1)Z}{4\pi c^2 Z^2}. \qquad (5.3.7)$$

To show (5.3.7) we proceed as follows:
(1) Using (5.3.5), the relation (5.2.4) gives $t_0 - t_s$ as a function of Z.
(2) We expand $A(r)$ as $A(r) = A(0) + r(dA/dr)_0$. The metric is not singular at $r = 0$ so $A(0)$ is non-zero; continuity at $r = 0$ of the partial derivatives of the metric gives $(dA/dr)_0 = 0$. The relation (5.2.2) then gives r_s as a function of $t_0 - t_s$.
(3) The results above yield r_s and $A(r_s)$ as functions of Z.
(4) Replacing r_s and $A(r_s)$ in (5.3.3) by these expressions and R_0/R_s by $1 + Z$, we obtain (5.3.6) corrected by the lowest-order term in Z, i.e. (5.3.7).

The luminosity L is estimated for a set of standard candles, while ℓ and Z are measured at Earth. Values of H_0 and q_0 are thus determined so as to fit (5.3.7) as well as possible.

We have already mentioned that current estimates of H_0 range from 40 km s^{-1}/Mpc to 100 km s^{-1}/Mpc. The value of q_0 deduced from (5.3.7) is uncertain (between 0.2 and 1.5, depending on the author). The use of observations in the context of various theoretical models gives information which is complementary, if at times contradictory: the parameter q_0 can be anywhere from a few per cent to a value of order unity.

Chapter 6

General relativity

We have already emphasised (cf. II § 2.4) the formal analogy between special-relativistic calculations in arbitrary coordinate systems and those in metric theories of gravitation. We shall make this analogy precise in this chapter and establish some useful formulae, without attempting a more profound treatment of the underlying framework; the latter is given explicitly in appendix B.

Once we have constructed the tools needed for calculations we shall return to a question we left unanswered: the determination of the function $A(r)$ in the cosmological models introduced in the previous chapter.

Finally we shall specify the gravitational field equations of general relativity.

6.1 Elements of tensor calculus

6.1.1 The notion of a tensor

The notion of a tensor introduced in special relativity carries through into the metric theories we now consider. We can recognise the components of a tensor by considering a coordinate change $x^{\alpha'} = x^{\alpha'}(x^{\beta})$, characterised by the coefficients $a^{\alpha'}{}_{\beta}(x) = \partial x^{\alpha'}/\partial x^{\beta}$. The components of a contravariant tensor field are written $A^{\alpha\beta\cdots}(x)$; they depend on the space–time point under consideration, and transform according to the rule

$$A^{\delta'\sigma'\cdots} = a^{\delta'}{}_{\alpha}a^{\sigma'}{}_{\beta\cdots} \cdot A^{\alpha\beta\cdots}. \qquad (6.1.1)$$

The transformation formulae for the components of a covariant tensor can be expressed using the coefficients $a^{\alpha}{}_{\beta'} = \partial x^{\alpha}/\partial x^{\beta'}$:

$$A_{\delta'\sigma'\cdots} = a^{\alpha}{}_{\delta'}a^{\beta}{}_{\sigma'} \cdot \ldots \cdot A_{\alpha\beta\cdots} \qquad (6.1.2)$$

We verify moreover the relations

$$a^{\alpha}{}_{\beta'}a^{\beta'}{}_{\tau} = \delta^{\alpha}{}_{\tau} \quad \text{and} \quad \delta^{\alpha'}{}_{\tau'} = a^{\alpha'}{}_{\beta}a^{\beta}{}_{\tau'}$$

where $\delta^{\alpha}{}_{\tau}$ and $\delta^{\alpha'}{}_{\tau'}$ are Kronecker symbols.

The metric tensor $g_{\alpha\beta}(x)$ is a symmetric covariant tensor (cf. II § 2.4.2); its inverse $g^{\alpha\beta}$ is symmetric and satisfies the following defining relations:

$$g^{\alpha\beta} g_{\alpha\sigma} = g_{\sigma\alpha} g^{\alpha\beta} = \delta^\beta{}_\sigma. \tag{6.1.3}$$

As an exercise, one can show that these relations, assumed to hold at all space–time points and for all coordinate systems, imply that $g^{\alpha\beta}$ are the components of a contravariant tensor.

The components of a mixed tensor have both upper and lower indices. In a coordinate change, the transformation law follows on using $a^\alpha{}_{\beta'}$ and $a^\alpha{}_\beta$ and balancing indices of the same type (cf. I (5.4.2)); for example

$$A^{\alpha'}{}_{\beta'} = a^{\alpha'}{}_\sigma a^\tau{}_{\beta'} A^\sigma{}_\tau. \tag{6.1.4}$$

As in special relativity (cf. I table 5.1), the tensors $g_{\alpha\beta}$ and $g^{\alpha\beta}$ provide a one–one correspondence between contravariant, covariant and mixed tensors, for example:

$$A^{\alpha\beta} \leftrightarrow A^\alpha{}_\beta = g_{\beta\sigma} A^{\alpha\sigma} \leftrightarrow A_{\alpha\beta} = g_{\alpha\tau} A^\tau{}_\beta \leftrightarrow A_\alpha{}^\beta = g_{\alpha\sigma} A^{\sigma\beta}$$
$$\text{with} \quad A^{\alpha\beta} = g^{\beta\sigma} A^\alpha{}_\sigma, \quad A^\alpha{}_\beta = g^{\alpha\sigma} A_{\sigma\beta} \dots$$

This correspondence allows us to regard $A_{\alpha\beta}$, $A^\alpha{}_\beta$, $A_\alpha{}^\beta$ and $A^{\alpha\beta}$ as the covariant, mixed and contravariant components of the same tensor. Indices are raised and lowered by means of contractions with $g^{\alpha\beta}$ and $g_{\alpha\beta}$ to pass from one type of components to a different type.

As an exercise the reader should show that this property is conserved under coordinate changes.

The order of a tensor is the number of indices characterising its components. Vectors are tensors of order one. Scalars, or tensors of order zero, are invariant under coordinate changes. The line element ds^2 and the mass of a particle are scalars for conceptual physical reasons; the trace $T^\alpha{}_\alpha$ of a mixed tensor is a scalar for mathematical reasons.

Consider a particle of a continuous fluid whose coordinates are parametrised using the proper time τ measured along the particle's world-line. As in special relativity we define the four-velocity with components U^α: $U^\alpha = dx^\alpha/d\tau$. Given an event with fixed coordinates $x = \{x^\alpha\}$ we consider the four-velocity of the particle at x; its components U^α are functions of x. We may verify that $U^\alpha(x)$ is a field of (four-)vectors obeying the relation $g_{\alpha\beta} U^\alpha U^\beta = c^2$.

6.1.2 Variation of a tensor

A gravitational field is characterised by the fact that there exists no coordinate system for which $g_{\alpha\beta} = \eta_{\alpha\beta}$ at every point. The absence of a privileged coordinate system immediately requires us to use general coordinates. To determine the variation of a tensor we have to use the relations established for the most general situations in special relativity (cf. I chapter 7); thus covariant derivatives appear in place of the partial derivatives appearing in special relativity in Minkowski coordinates.

Vectors are defined locally at each point of space-time in a vector space called the **tangent space** at the point. The tangent spaces at different points differ from each other. It is therefore not trivial to define the variation of a vector in passing from one space-time point to a neighbouring one over time for example. For a discussion of the extension of the geometrical framework of special relativity to relativistic metric theories of gravity the reader should consult appendix B. The current discussion deliberately aims at giving a prescription rather than a theoretical exposition.

We recall an example of the covariant derivative of a field of tensors:

$$\nabla_\mu T^\alpha{}_\beta = \partial_\mu T^\alpha{}_\beta + \Gamma^\alpha{}_{\sigma\mu} T^\sigma{}_\beta - \Gamma^\sigma{}_{\beta\mu} T^\alpha{}_\sigma \qquad (6.1.5)$$

where the connection coefficients $\Gamma^\alpha{}_{\beta\sigma}$ are equal to the Christoffel symbols, $\{^\alpha{}_{\beta\sigma}\}$:

$$\Gamma^\alpha{}_{\beta\sigma} = \{^\alpha{}_{\beta\sigma}\} \equiv \frac{1}{2} g^{\alpha\tau}(g_{\beta\tau,\sigma} + g_{\sigma\tau,\beta} - g_{\beta\sigma,\tau}). \qquad (6.1.6)$$

In the most general theories, the connection, $\Gamma^\alpha{}_{\beta\sigma}$, is not necessarily equal to the Christoffel symbol $\{^\alpha{}_{\beta\sigma}\}$. This equality is, however, assured for metric theories based on the Einstein equivalence principle (cf. appendix B). In the following we shall assume this equality which we regard as a basic postulate of general relativity.

We may verify Leibniz's rule, of which we give an example:

$$\nabla_\nu[A^\alpha{}_\beta \cdot B^{\sigma\mu}] = (\nabla_\nu[A^\alpha{}_\beta]) \cdot (B^{\sigma\mu}) + (A^\alpha{}_\beta) \cdot (\nabla_\nu[B^{\sigma\mu}]). \qquad (6.1.7)$$

Direct calculation shows that the covariant derivative of the metric tensor vanishes:

$$\nabla_\mu g_{\alpha\beta} = 0 \quad \text{and} \quad \nabla_\mu g^{\alpha\beta} = 0. \qquad (6.1.8)$$

From these two rules it is easy to show that indices can be raised and lowered either before or after covariant differentiation; for example

$$g_{\alpha\beta} \nabla_\mu T^{\beta\cdots} = \nabla_\mu T_\alpha{}^{\cdots}. \qquad (6.1.9)$$

6.1.3 Minimal coupling

We can easily show that the covariant derivative of a tensor is a tensor. This provides a justification for using covariant derivatives in the formulation of physical laws.

Consider for example Maxwell's equations in vacuo (I 6.2.1).

In appendix B we show how to determine a coordinate system such that at an arbitrarily chosen space-time point A the Christoffel symbols vanish and $g_{\alpha\beta} = \eta_{\alpha\beta}$. This system can be identified with a Minkowski system of special relativity in the neighbourhood of A. The Einstein equivalence principle suggests that in this coordinate system Maxwell's equations at A should be written

$$(\partial_\beta F^{\alpha\beta})_A = 0 \quad \text{and} \quad (\partial_\tau F_{\alpha\beta} + \partial_\beta F_{\tau\alpha} + \partial_\alpha F_{\beta\tau})_A = 0,$$
$$\text{with} \quad (F_{\alpha\beta})_A = (\eta_{\alpha\sigma}\eta_{\beta\tau}F^{\sigma\tau})_A.$$

In this coordinate system, where the connection coefficients vanish at A, these equations can be rewritten

$$(\nabla_\beta F^{\alpha\beta})_A = 0 \quad \text{and} \quad (\nabla_\tau F_{\alpha\beta} + \nabla_\beta F_{\tau\alpha} + \nabla_\alpha F_{\beta\tau})_A = 0,$$
$$\text{with} \quad (F_{\alpha\beta})_A = (g_{\alpha\sigma}g_{\beta\tau}F^{\sigma\tau})_A.$$

We assume that $F^{\alpha\beta}$ is a tensor. This postulate implies certain well-verified physical properties, such as the transformations of the electric and magnetic fields for two observers moving with respect to each other. There is no reason to doubt its validity. Under these conditions, Maxwell's equations at A imply the vanishing of the tensors $(\nabla_\beta F^{\alpha\beta})_A$ and $(\nabla_\tau F_{\alpha\beta} + \nabla_\beta F_{\tau\alpha} + \nabla_\alpha F_{\beta\tau})_A$. As the point A was arbitrary, we do not need to specify it. The vanishing of these tensors can moreover be expressed in an arbitrary coordinate system. Quite generally, then, Maxwell's equations in vacuo are

$$\nabla_\beta F^{\alpha\beta} = 0 \text{ and } \nabla_\tau F_{\alpha\beta} + \nabla_\beta F_{\tau\alpha} + \nabla_\alpha F_{\beta\tau} = 0, \text{ with } F_{\alpha\beta} = g_{\alpha\sigma}g_{\beta\tau}F^{\sigma\tau}.$$

Verify that the vanishing of the components of a tensor in a particular coordinate system implies that they vanish in any system. Thus the vanishing of a tensor, i.e. of all of its components, is a property of the tensor and not of the coordinate system used.

We shall denote as **minimal coupling** this manner of introducing gravitation through the substitution of $g_{\alpha\beta}$ for $\eta_{\alpha\beta}$ and of ∇_β for ∂_β in equations written in special relativity in Minkowski coordinates.

This procedure, suggested by the Einstein equivalence principle, is not the only possible way of generalising. It is certainly the simplest, and there is currently no experimental evidence throwing doubt upon it.

When a physical law is written in the form

Tensor [function of those fields present] $= \mathbf{0}$

we say that *the law is written in "covariant" form.* Maxwell's equations in vacuo as written above thus express the "coupling" of electromagnetism with gravitation in covariant form.

6.1.4 Important tensor identities

In special relativity the operators ∇_μ and ∇_σ commute:

$$\nabla_\mu \nabla_\sigma A^{\alpha\cdots}{}_{\beta\ldots} - \nabla_\sigma \nabla_\mu A^{\alpha\cdots}{}_{\beta\ldots} \equiv [\nabla_\mu, \nabla_\sigma] A^{\alpha\cdots}{}_{\beta\ldots} \equiv 0.$$

In general relativity the situation is different. As an exercise the reader can prove the relation

$$[\nabla_\mu, \nabla_\sigma] A^{\alpha\cdots}{}_{\beta\ldots} \equiv -R^\alpha{}_{\varepsilon\mu\sigma} A^{\varepsilon\cdots}{}_{\beta\ldots} - \ldots + R^\varepsilon{}_{\beta\mu\sigma} A^{\alpha\cdots}{}_{\varepsilon\ldots} + \ldots \tag{6.1.10}$$

with

♦
$$R^\alpha{}_{\beta\mu\sigma} \equiv \Gamma^\alpha{}_{\beta\mu,\sigma} - \Gamma^\alpha{}_{\beta\sigma,\mu} + \Gamma^\alpha{}_{\varepsilon\sigma}\Gamma^\varepsilon{}_{\beta\mu} - \Gamma^\alpha{}_{\varepsilon\mu}\Gamma^\varepsilon{}_{\beta\sigma}. \tag{6.1.11}$$

$R^\alpha{}_{\beta\mu\sigma}$ is the **curvature** or **Riemann–Christoffel tensor**, usually called simply the Riemann tensor. In appendix B we show that the absence of a gravitational field is equivalent to the vanishing of the Riemann tensor.

Contraction of the Riemann tensor gives the **Ricci tensor**

♦
$$R_{\beta\mu} \equiv R^\alpha{}_{\beta\mu\alpha}. \tag{6.1.12}$$

This tensor is symmetric: $R_{\beta\mu} \equiv R_{\mu\beta}$.

The **scalar curvature** is given by contraction of the Ricci tensor:

♦
$$R = R^\alpha{}_\alpha. \tag{6.1.13}$$

Also used is the **Einstein tensor**:

♦
$$S_{\alpha\beta} \equiv R_{\alpha\beta} - \frac{1}{2} g_{\alpha\beta} R. \tag{6.1.14}$$

This tensor satisfies the following identity, proved in appendix B:

♦
$$S^{\alpha\beta}{}_{;\beta} \equiv 0 \tag{6.1.15}$$

where we have used the notation $(\ )_{;\mu} \equiv \nabla_\mu(\ )$.

6.2 Parallel transport

In the present context the physical content of space–time is described by tensor fields.

Consider the physical state in the neighbourhood of a point with coordinates x; to reproduce the same physical state at x' we must "suitably" transport there the physical systems initially at x along some space-time path. The mathematical equivalent of this operation is called **parallel transport**.

Given a tensor field $T^\alpha{}_\beta$ for example, we use the covariant derivative to determine the change $\nabla T^\alpha{}_\beta$ of the tensor between two neighbouring points with coordinates $\{x^\mu\}$ and $\{x^\mu + dx^\mu\}$:

$$\nabla T^\alpha{}_\beta \equiv (\nabla_\mu T^\alpha{}_\beta) dx^\mu \equiv (\partial_\mu T^\alpha{}_\beta + \Gamma^\alpha{}_{\sigma\mu} T^\sigma{}_\beta - \Gamma^\sigma{}_{\beta\mu} T^\alpha{}_\sigma) dx^\mu. \tag{6.2.1}$$

As in special relativity, the variation $\nabla T^{\alpha}{}_{\beta}$ of a tensor takes account of the variation $dT^{\alpha}{}_{\beta}$ of the tensor components and that of the basis vectors of the tangent spaces over which the decomposition is performed (appendix B).

If the change of the tensor considered vanishes between $\{x^{\mu}\}$ and $\{x^{\mu}+dx^{\mu}\}$ we say that the tensor has been parallel transported. The variation of the components of $T^{\alpha}{}_{\beta}$ is then

$$dT^{\alpha}{}_{\beta} \equiv \partial_{\mu}T^{\alpha}{}_{\beta}dx^{\mu} = (-\Gamma^{\alpha}{}_{\sigma\mu}T^{\sigma}{}_{\beta} + \Gamma^{\sigma}{}_{\beta\mu}T^{\alpha}{}_{\sigma})dx^{\mu}. \tag{6.2.2}$$

Consider an "isolated" tensor defined at $\{x^{\mu}\}$, for example $F^{\alpha}{}_{\beta}$. We do not consider a field of tensors; however, in the parallel transport of the tensor from $\{x^{\mu}\}$ to $\{x^{\mu} + dx^{\mu}\}$ its components will change according to the second equality of (6.2.2). Parallel transport thus gives at $\{x^{\mu} + dx^{\mu}\}$ the tensor with components $F^{\alpha}{}_{\beta} + (-\Gamma^{\alpha}{}_{\sigma\mu}F^{\sigma}{}_{\beta} + \Gamma^{\sigma}{}_{\beta\mu}F^{\alpha}{}_{\sigma})dx^{\mu}$. By integrating we may thus define the parallel transport of a tensor along a path from x to x'. If there is curvature, the resulting tensor will in general depend on the path followed between the two points with coordinates x and x'.

As an exercise show that a unit vector tangent to a geodesic remains a unit vector tangent to the geodesic under parallel transport along it. In this sense a geodesic is a **self-parallel** curve. Show further that the scalar product of two vectors is preserved under parallel transport.

6.3 Isotropy in cosmology

We can now specify how the function $A(r)$ introduced in the last chapter is calculated in homogeneous and isotropic cosmological models.

Consider a space-time with line element (5.1.1):

$$ds^2 = c^2 dt^2 - R^2(t)A^2(r)(dx^2 + dy^2 + dz^2). \tag{6.3.1}$$

Let $\{V^{\alpha}\}$ and $\{W^{\alpha}\}$ be two spacelike unit vectors with $V^0 = W^0 = 0$ and consider a second rank tensor constructed purely from the metric coefficients, for example the Ricci tensor $R_{\alpha\beta}$. The condition of isotropy at each point implies that $R_{\alpha\beta}V^{\alpha}V^{\beta} = R_{\alpha\beta}W^{\alpha}W^{\beta}$. Choosing $\{V^{\alpha}\} = \{0,(AR)^{-1},0,0\}$ and $\{W^{\alpha}\} = \{0,0,(AR)^{-1},0\}$ we find $R_{11} = R_{22}$. Expressing $R_{\alpha\beta}$ in terms of the metric, we find a differential equation satisfied by the function $A(r)$ whose solution gives the expression (5.1.2):

$$A(r) = \frac{1}{1 + \frac{1}{4}kr^2} \quad \text{with} \quad k = 1, -1 \text{ or } 0. \tag{6.3.2}$$

The differential equation is $-\dfrac{d}{dr}\left(\dfrac{1}{rA}\dfrac{dA}{dr}\right) + r\left(\dfrac{1}{rA}\dfrac{dA}{dr}\right)^2 = 0.$

We can easily verify that the solution has the form $A(r) = \dfrac{a}{1 + br^2}$ where a and b are two integration constants.

For $b = 0$, we introduce a function $R'(t) = |a| \cdot R(t)$. The line element is then (6.3.1) up to notation, with $k = 0$ in (6.3.2).

For $b \neq 0$ we set $b = k \cdot |b|$, with $k = 1$ or -1 according to the sign of b. We introduce coordinates $x^{j'} = x^j \sqrt{|4b|}$ so that the function $R'(t) = |a|R(t)/\sqrt{|4b|}$. The line element then takes the form (6.3.1) up to notation, with $k = 1$ or -1 according to the sign of b.

We note that r, x^j and A are dimensionless while $R(t)$ is a length.

In spherical coordinates the line element (6.3.1) is

$$ds^2 = c^2 dt^2 - \frac{R^2(t)}{\left(1 + \dfrac{kr^2}{4}\right)^2} (dr^2 + r^2 \sin^2\theta\, d\phi^2 + r^2 d\theta^2). \qquad (6.3.3)$$

Changing the radial coordinate alone, we can put the line element in the form

$$\blacklozenge \qquad ds^2 = c^2 dt^2 - R^2(t) \left(\frac{d\bar{r}^2}{1 - k\bar{r}^2} + \bar{r}^2 \sin^2\theta\, d\phi^2 + \bar{r}^2\, d\theta^2 \right) \qquad (6.3.4)$$

with

$$\bar{r} = r \left| 1 + \frac{kr^2}{4} \right|^{-1}. \qquad (6.3.5)$$

Expressions (6.3.3) and (6.3.4) are locally equivalent, although they are not globally equivalent as they cover different sets of events. For $k = 1$ both of the two regions $0 < r < 2$ and $2 < r < \infty$ correspond to $0 < \bar{r} < 1$. For $k = -1$ these two regions correspond to the interval $\bar{r} \in [0, \infty[$. Thus in each case the metric (6.3.4) describes one of the two regions only.

The assumptions adopted lead necessarily to the line element (6.3.4). We still have to check that the line element indeed represents a homogeneous and isotropic space-time.

To do this we wish to show the existence of a coordinate system $\{t, \bar{r}', \theta', \phi'\}$ such that $\bar{r}' = 0$ is a point with initial coordinates $\{\bar{r}, \theta, \phi\}$ chosen arbitrarily, with the metric formally unchanged:

$$ds^2 = c^2 dt^2 - R^2(t) \left(\frac{d\bar{r}'^2}{1 - k\bar{r}'^2} + \bar{r}'^2 \sin^2\theta' d\phi'^2 + \bar{r}'^2 d\theta'^2 \right). \qquad (6.3.6)$$

To prove this we embed the hypersurface E_t with equation $t = $ constant in a suitably chosen auxiliary four-dimensional space. The space E_t then appears as a three-dimensional surface of this auxiliary space with the appropriate properties.

The various metrics (6.3.4) are compatible with differing topologies for E_t according to the value of k:

$k = 1$, $\bar{r} \in [0, 1[$, $\theta \in [0, \pi]$, $\phi \in [0, 2\pi[$. The space E_t is spherical and closed. Its volume is $\pi^2 R^3(t)$ (cf. § 5.3.1, the calculation of V with $r_0 = 2$)

$k = -1$, $\bar{r} \in [0, \infty[$, $\theta \in [0, \pi]$, $\phi \in [0, 2\pi[$. The space is hyperbolic and open. Its volume is infinite

$k = 0$, $\bar{r} \in [0, \infty[$, $\theta \in [0, \pi]$, $\phi \in [0, 2\pi[$. The space E_t is Euclidean.

For $k = 1$ we set $\bar{r} = \sin \Omega$. The line element (6.3.4) becomes $ds^2 = c^2 dt^2 - d\sigma^2$ where $d\sigma$ is the element of proper length in E_t:

$$d\sigma^2 = R^2(t) (d\Omega^2 + \sin^2\Omega \, d\theta^2 + \sin^2\Omega \sin^2\theta \, d\phi^2).$$

To study E_t we consider a Euclidean space of four dimensions, with cartesian coordinates x, y, z, u. We introduce spherical coordinates R, Ω, θ, ϕ:

$$x = R \sin \Omega \sin \theta \cos \phi \quad y = R \sin \Omega \sin \theta \sin \phi$$
$$z = R \sin \Omega \cos \theta \quad u = R \cos \Omega.$$

The element of length $d\Sigma$ in this space is

$$d\Sigma^2 = dx^2 + dy^2 + dz^2 + du^2 = R^2(d\Omega^2 + \sin^2\Omega \, d\theta^2 + \sin^2\Omega \sin^2\theta \, d\phi^2) + dR^2.$$

E_t is a sphere centred at the origin, of radius $R = R(t)$ and with equation $x^2 + y^2 + z^2 + u^2 = R^2$. The element of length is just $d\sigma$ on this sphere. An arbitrary point A of E_t is represented by a point on the sphere. We make a rotation of the coordinate axes such that the u-axis passes through A. The equation of the sphere remains formally unchanged, as do the expressions for $d\Sigma$ and $d\sigma$:

$$x'^2 + y'^2 + z'^2 + u'^2 = R^2(t)$$
$$d\Sigma^2 = R^2 (d\Omega'^2 + \sin^2\Omega' \, d\theta'^2 + \sin^2\Omega' \sin^2\theta' \, d\phi'^2) + dR^2$$
$$d\sigma^2 = R^2(t) (d\Omega'^2 + \sin^2\Omega' \, d\theta'^2 + \sin^2\Omega' \sin^2\theta' \, d\phi'^2).$$

The line element is thus formally preserved under the coordinate change considered. Setting $\bar{r}' = \sin \Omega'$ we get

$$ds^2 = c^2 dt^2 - R^2(t) \left(\frac{d\bar{r}'^2}{1 - \bar{r}'^2} + \bar{r}'^2 \sin^2\theta' d\phi'^2 + \bar{r}'^2 d\theta'^2 \right).$$

Point A, whose coordinates t, Ω, θ and ϕ are arbitrary, is characterised in the new coordinate system by the value $\Omega' = 0$, i.e. $\bar{r}' = 0$. We have thus shown that the form (6.3.6) of the line element is compatible for $k = 1$ with the choice of an origin A of arbitrary initial spatial coordinates.

For $k = -1$ we set $\bar{r} = \sinh \Omega$. The line element (6.3.4) is $ds^2 = c^2 dt^2 - d\sigma^2$ where $d\sigma$ is the element of proper length in E_t:

$$d\sigma^2 = R^2(t) (d\Omega^2 + \sinh^2\Omega \, d\theta^2 + \sinh^2\Omega \sin^2\theta \, d\phi^2).$$

Consider a four-dimensional pseudo-Euclidean space with Minkowski coordinates x, y, z, u. We introduce the coordinates R, Ω, θ, ϕ:

$$x = R \sinh \Omega \sin \theta \cos \phi \quad y = R \sinh \Omega \sin \theta \sin \phi$$
$$z = R \sinh \Omega \cos \theta \quad u = R \cosh \Omega.$$

The line element becomes

$$d\Sigma^2 = du^2 - dx^2 - dy^2 - dz^2 = -R^2 (d\Omega^2 + \sinh^2\Omega \, d\theta^2 + \sinh^2\Omega \sin^2\theta \, d\phi^2) + dR^2.$$

E_t now appears as a hyperboloid of revolution with respect to the u-axis, with equation $u^2 - x^2 - y^2 - z^2 = R^2 = R^2(t)$. The element of length is just $d\sigma$ on this hyperboloid. An arbitrary point A of E_t is represented by a point on the hyperboloid. We make a Lorentz rotation of the coordinate axes such that the u-axis passes through A. The formulae for coordinate changes are then Lorentz transformations in the auxiliary space that we have introduced. The equation of the hyperboloid remains formally unchanged, as do the expressions for $d\Sigma$ and $d\sigma$:

$$u'^2 - x'^2 - y'^2 - z'^2 = R^2(t)$$
$$d\Sigma^2 = -R^2(d\Omega'^2 + \sinh^2\Omega' d\theta'^2 + \sinh^2\Omega' \sin^2\theta' d\phi'^2) + dR^2$$
$$d\sigma^2 = R^2(t)(d\Omega'^2 + \sinh^2\Omega' d\theta'^2 + \sinh^2\Omega' \sin^2\theta' d\phi'^2).$$

The line element is thus formally preserved under the coordinate change considered.

Setting $\bar{r}' = \sinh \Omega'$, we get

$$ds^2 = c^2 dt^2 - R^2(t) \left(\frac{d\bar{r}'^2}{1 + \bar{r}'^2} + \bar{r}'^2 \sin^2\theta' d\phi'^2 + \bar{r}'^2 d\theta'^2 \right).$$

The point A, whose coordinates t, Ω, θ and ϕ are arbitrary, is characterised in the new coordinate system by the value $\Omega' = 0$, i.e. $\bar{r}' = 0$. We have thus shown that the form (6.3.6) of the line element is compatible, for $k = -1$, with the choice of an origin, A, of arbitrary spatial coordinates.

The case $\mathbf{k} = 0$ is simpler. We use coordinates $\{x, y, z\}$ defined by:

$$x = \bar{r} \sin \theta \cos \phi, \quad y = \bar{r} \sin \theta \sin \phi, \quad z = \bar{r} \cos \theta.$$

The element of proper length in E_t obeys

$$d\sigma^2 = R^2(t) \, (dx^2 + dy^2 + dz^2).$$

A coordinate translation allows us to choose an arbitrary point A of E_t as origin without changing the form of $d\sigma^2$. Changing to spherical coordinates with this point as origin then gives the line element (6.3.6).

In introducing the metrics (6.3.1) we assumed in the last chapter that space is homogeneous and isotropic. In reality it is enough to assume that space is isotropic at each point: homogeneity then follows.

6.4 Einstein's equations

In Newtonian theory the gravitational field $\vec{g}$ is found from the gravitational potential U; the latter is related to the matter distribution in the universe through the matter density ρ (cf. II § 1.3.1):

$$\vec{g} = -\overrightarrow{\mathrm{grad}}[U] \qquad \Delta U = 4\pi G\rho. \qquad (6.4.1)$$

We also postulate *independently* the conservation of mass and Newton's second law (II 4.4.1 and 4.4.2):

$$\frac{\partial \rho}{\partial t} + \mathrm{div}[\rho \vec{V}] = 0 \qquad \frac{\partial \vec{V}}{\partial t} + (\vec{V} \cdot \vec{\nabla})\, \vec{V} = \vec{g} \qquad (6.4.2)$$

where $\vec{V}$ is the velocity field of the fluid considered.

In special relativity there is no gravitational field. Matter is described by its energy–momentum tensor $T^{\alpha\beta} = T^{\beta\alpha}$ (cf. I chapter 6). In a Minkowski system the relativistic equivalent of the two relations (6.4.2) is

$$\partial_\alpha T^{\alpha\beta} = 0. \qquad (6.4.3)$$

In general relativity our earlier analysis (§ 6.1.3) shows that this equation becomes:

♦
$$\nabla_\alpha T^{\alpha\beta} \equiv T^{\alpha\beta}{}_{;\alpha} = 0. \qquad (6.4.4)$$

$T^{\alpha\beta}$ is the generalisation of the corresponding tensor in special relativity; for a perfect fluid

♦
$$T^{\alpha\beta} = \left(\rho + \frac{P}{c^2}\right) U^\alpha U^\beta - P g^{\alpha\beta}$$

ρ is the mass density in the tangent reference frame, P the pressure and U^α the four-velocity field.

Equations (6.4.4) are regarded as both "conservation equations" and "equations of motion".

> We can introduce a freely falling observer who accompanies a given fluid particle for an infinitesimal time interval. We obtain the earlier expression for $T^{\alpha\beta}$ by operating as for Maxwell's equations in section 6.1.3. We may thus deduce the physical significance of ρ and P from their interpretation in special relativity. We may check that $T^{\alpha\beta}$ is a tensor.

In general relativity, gravitation is described by the metric tensor and the matter distribution by the tensor $T^{\alpha\beta}$. The field equations of gravity are equations relating the metric tensor and the energy–momentum tensor. They are postulated in the manifestly covariant form

♦
$$S^{\alpha\beta} = \chi \cdot T^{\alpha\beta} \qquad (6.4.5)$$

where χ is a constant and $S^{\alpha\beta}$ the Einstein tensor defined above (6.1.14).

These are **Einstein's equations**.

The identities (6.1.15), $S^{\alpha\beta}{}_{;\beta} \equiv 0$, then imply the conservation equations (6.4.4), without the necessity of postulating them separately. This great conceptual economy is an attractive feature of general relativity as a physical theory.

The energy–momentum tensor describes not only matter distributions but also fields of all kinds, such as the electromagnetic field for example. However, the right hand side of Einstein's equations (6.4.5) excludes contributions of purely gravitational origin.

Einstein's equations can also be written in the form

$$R^{\alpha\beta} = \chi \cdot \left(T^{\alpha\beta} - \frac{1}{2} g^{\alpha\beta} T \right) \quad \text{with} \quad T = T^{\alpha}{}_{\alpha}. \tag{6.4.6}$$

Contracting (6.1.14) we find $S^{\alpha}{}_{\alpha} = -R$. Contraction of (6.4.5) gives $S^{\alpha}{}_{\alpha} = \chi \cdot T$. Einstein's equations then become

$$S^{\alpha\beta} \equiv R^{\alpha\beta} - \frac{1}{2} g^{\alpha\beta} R = \chi \cdot T^{\alpha\beta} \Rightarrow R^{\alpha\beta} = \chi \cdot T^{\alpha\beta} + \frac{1}{2} g^{\alpha\beta} R = \chi \cdot \left(T^{\alpha\beta} - \frac{1}{2} g^{\alpha\beta} T \right).$$

Conversely one can show that (6.4.6) implies (6.4.5).

The Einstein tensor is not the only one satisfying an identity of the type (6.1.15). Other theories have been conceived in which the field equations have the form $E^{\alpha\beta} = \chi \cdot T^{\alpha\beta}$ with $E^{\alpha\beta}{}_{;\beta} \equiv 0$. However, if we require that the tensor $E^{\alpha\beta}$ should contain derivatives of $g_{\alpha\beta}$ of order no higher than two and that the second derivatives appear linearly, the choice is restricted:

$$E^{\alpha\beta} \equiv A \cdot (S^{\alpha\beta} - \Lambda \cdot g^{\alpha\beta})$$

where A and Λ are constants. Λ is called the **cosmological constant** as it was introduced by Einstein in a static cosmological model before the discovery of the Hubble expansion. We shall assume it to be zero in the following, which is compatible with current observations. Under these conditions, Einstein's equations are the only ones possible.

We denote as **general relativity** the metric theory having Einstein's equations as the field equations of gravitation.

6.5 Newtonian limit of general relativity

Here we shall show that general relativity reduces to Newtonian theory in its domain of validity, i.e. for weak gravitational fields and small velocities.

As the field is weak we may use a quasi-Minkowski coordinate system, which we can identify with a quasi-Galilean system in Newtonian theory: $g_{\alpha\beta} = \eta_{\alpha\beta} + h_{\alpha\beta}$, with $|h_{\alpha\beta}| \ll 1$.

We consider only the case where the gravitational field is static, $g_{\alpha\beta,0} = 0$. Under these conditions, with the velocities with respect to the coordinate system remaining small, we recover Newtonian theory by setting

$$g_{00} = 1 + \frac{2U}{c^2} \tag{6.5.1}$$

where U is the Newtonian gravitational potential (cf. II § 2.4.2). The other components $g_{\alpha\beta}$ are arbitrary, although close to their Minkowski values.

Neglecting terms quadratic in $|h_{\alpha\beta}|$ and using the fact that the gravitational field is static, R_{00} can be expressed as a function of g_{00} alone:

$$R^{00} \approx R_{00} \approx \Gamma^{\alpha}{}_{00,\alpha} = \left(\frac{1}{2}\,g^{\alpha\sigma}(-g_{00,\sigma})\right)_{,\alpha} \approx \frac{1}{2}\,\Delta g_{00} = \frac{\Delta U}{c^2}. \qquad (6.5.2)$$

To lowest order the energy–momentum tensor takes its Minkowski form; for an incoherent fluid with $P = 0$ we have

$$T^{00} = \rho(U^0)^2 \approx \rho c^2, \; T^{0k} = \rho U^0 U^k \approx \rho c V^k,$$
$$T^{kj} = \rho U^k U^j \approx \rho V^k V^j \qquad (6.5.3)$$

where V^k is the Newtonian velocity field of the fluid. To lowest order in $|V^k/c|$ we find

$$T^{00} \approx \rho c^2 \approx T. \qquad (6.5.4)$$

The ^{00} component of Einstein's equations in the form (6.4.6) gives the Newtonian equation

$$\Delta U = 4\pi G \rho \qquad (6.5.5)$$

if we set

♦
$$\chi \equiv \frac{8\pi G}{c^4}. \qquad (6.5.6)$$

General relativity thus includes Newtonian theory as a limiting case when the gravitational field is weak, provided that we do not need to consider the propagation of the field in time, and that velocities are small.

6.6 Geodesics and equations of motion

Consider an incoherent fluid consisting of particles with very small masses, subject to the gravitational field produced by itself and any other sources. Every particle therefore behaves like a test particle in free fall.

Einstein's equations imply the equations of motion (6.4.4) for the fluid: $T^{\alpha\beta}{}_{;\beta} = 0$. However, from our basic postulates we know that a test particle in free fall moves along a geodesic (II § 2.3). There is no contradiction between these two ways of determining the motion as equations (6.4.4) also imply that freely falling particles move on geodesics. We recall that one of the consequences of this property is the identity of the free-fall motion for particles with the same initial conditions. We explained this property in a Newtonian framework using the equivalence principle.

Thus Einstein's equations and the associated interpretation appear able to provide a relativistic theory of gravitation which is in agreement with Newtonian theory in its domain of validity.

Consider an incoherent fluid with the energy–momentum tensor given above as (6.5.3). The equations of motion can be written

$$T^{\alpha\beta}{}_{;\beta} \equiv (\rho U^{\alpha} U^{\beta})_{;\beta} \equiv (\rho U^{\beta})_{;\beta} U^{\alpha} + \rho U^{\beta}(U^{\alpha})_{;\beta} = 0.$$

Further, $U^{\alpha}U_{\alpha} = c^2 \Rightarrow (U^{\alpha}U_{\alpha})_{;\beta} \equiv 2U_{\alpha}U^{\alpha}{}_{;\beta} = 0 \Rightarrow U_{\alpha}T^{\alpha\beta}{}_{;\beta} = c^2(\rho U^{\beta})_{;\beta} = 0.$
Thus the equations of motion imply

$$U^{\beta}U^{\alpha}{}_{;\beta} \equiv U^{\beta}\left(\partial_{\beta}U^{\alpha} + \Gamma^{\alpha}{}_{\sigma\beta}U^{\sigma}\right) = 0.$$

Following the motion of a particle on its worldline, parametrised by the proper time τ: $x^{\alpha} = X^{\alpha}(\tau)$, $U^{\alpha} = \dfrac{dX^{\alpha}}{d\tau} \Rightarrow U^{\beta}\partial_{\beta}U^{\alpha} = \dfrac{dU^{\alpha}}{d\tau}$.
We find the equations of motion:

$$U^{\beta}U^{\alpha}{}_{;\beta} \equiv \frac{d^2 X^{\alpha}}{d\tau^2} + \Gamma^{\alpha}{}_{\sigma\beta}\frac{dX^{\sigma}}{d\tau}\frac{dX^{\beta}}{d\tau} = 0.$$

These equations are exactly those of a geodesic (cf. II 2.3.5).

Chapter 7

Applications of Einstein's equations

In this chapter we shall use the field equations in two applications of general relativity, to cosmology and to spherically symmetric gravitational fields.

7.1 Einstein's equations in cosmology

Consider the cosmological spaces we have already studied in chapter 5 of part II. These have the line element (6.3.4)

$$ds^2 = c^2 dt^2 - R^2(t) \left(\frac{dr^2}{1 - kr^2} + r^2 \sin^2 \theta \, d\phi^2 + r^2 d\theta^2 \right) \qquad (7.1.1)$$

with $k = 1, 0$ or -1.

We model the contents of the Universe as a perfect fluid of pressure P and density ρ. The energy–momentum tensor is

$$T^{\alpha\beta} = \left(\rho + \frac{P}{c^2} \right) U^\alpha U^\beta - P g^{\alpha\beta} \qquad (7.1.2)$$

where U^α is the four-velocity of a particle of the cosmological fluid (cf. II chapter 6). As the coordinate system is comoving with the cosmological fluid we have $\{U^\alpha\} = \{c, 0, 0, 0\}$.

Einstein's equations thus reduce to just the two equations

$$\begin{cases} 3\ddot{R} = -4\pi G \left(\rho + \frac{3P}{c^2} \right) R \\ R\ddot{R} + 2\dot{R}^2 + 2kc^2 = 4\pi G \left(\rho - \frac{P}{c^2} \right) R^2 \end{cases} \qquad (7.1.3)$$

where G is the Newtonian gravitational constant. We have also used the notation $(\ ) \equiv d(\ )/dt \equiv c(\ )_{,0}$.

As R is a function of t alone, equations (7.1.3) imply that ρ and P are also functions of t only, in agreement with spatial homogeneity.

The condition of isotropy shows that the only mixed non-zero components of the Ricci tensor resulting from the metric (7.1.1) are $R^0_{\ 0}$ and $R^r_{\ r} = R^\theta_{\ \theta} = R^\phi_{\ \phi}$. This result can be obtained by generalising the method used in section 6.3 for arbitrary spacelike unit vectors, V^α and W^α. An explicit calculation leads to the expressions

$$R_{00} = -\frac{3\ddot{R}}{c^2 R} \qquad R_{\theta\theta} = r^2 \left(\frac{R\ddot{R}}{c^2} + \frac{2\dot{R}^2}{c^2} + 2k \right).$$

As the coordinate system is comoving with the cosmological fluid, the four–velocity field has components $U^0 = c$, $U^r = U^\theta = U^\phi = 0$. Explicit calculation shows that Einstein's equations in the form (6.4.6) reduce to the two equations (7.1.3).

Eliminating $\ddot{R}$ between the two equations (7.1.3) we find the relation

$$\dot{R}^2 + kc^2 = \frac{8\pi G}{3} \rho R^2. \qquad (7.1.4)$$

At the present epoch, $t = 0$, this equation may be regarded as a numerical relation between the Hubble constant $H_0 = (\dot{R}/R)_0$, the density, ρ_0, and k. We set $\rho_c = 3H_0^2/8\pi G$. Then (7.1.4) shows that the sign of k is that of $\rho_0 - \rho_c$. The possible values of k are 1, 0 and -1, giving the following:

$$\rho_c \equiv \frac{3H_0^2}{8\pi G},$$

$$\rho_0 = \rho_c \Rightarrow k = 0$$

$$\rho_0 < \rho_c \Rightarrow k = -1 : \text{ the model is hyperbolic}$$

$$\rho_0 > \rho_c \Rightarrow k = 1 : \text{ the model is spherical.}$$

It is interesting to note the correspondence between these results and the Newtonian ones (II § 4.6).

Similarly, the first of the two relations (7.1.3) gives an expression for the deceleration parameter q_0 defined earlier (II § 5.3.3 in relativity and II § 4.6 in Newtonian theory),

$$q_0 \equiv -\frac{\ddot{R}R}{\dot{R}^2} = \frac{4\pi G}{3H_0^2} \left(\rho_0 + \frac{3P_0}{c^2} \right) \approx \frac{1}{2} \frac{\rho_0}{\rho_c}.$$

We may assume that the current pressure P_0 is negligible compared with $\rho_0 c^2$; we thus get $\rho_0 = 2q_0\rho_c$ (cf. 4.6.2).

The values of H_0 and q_0 may be determined from observation (cf. II § 5.3.3). We deduce the values of ρ_c and ρ_0. The resulting value of ρ_0 is very uncertain, but much larger than that found by direct observation of visible matter. We have already mentioned this problem of "missing mass" (cf. II § 4.6).

> The value of H_0 is of order 75 km s^{-1}/Mpc, with a dispersion of about 30% depending on the authors. The value of q_0 is of order unity (cf. § 5.3.3).
> The observed density is of order 3×10^{-31} g/cm$^3 = 3 \times 10^{-28}$ kg/m^3, while the density deduced from the values just quoted is about one hundred times larger, i.e. 2×10^{-29} g/cm^3, or close to twice the critical density for $q_0 = 1$.
> The problem of missing mass also appears in the study of stars orbiting in galaxies. Assuming that all the matter is visible, the orbital velocity should decrease with the distance from the centre of the galaxy more rapidly than observed.

From equations (7.1.3) we deduce the relation

$$\frac{d}{dR}(\rho R^3) = -\frac{3P}{c^2}R^2. \tag{7.1.5}$$

In this expression ρ and P are functions of R. Giving an equation of state in the form $P = P(\rho)$ allows us to find ρ and P as functions of R.

Differentiating equation (7.1.4) and eliminating $\ddot{R}$ between the result and the first of equations (7.1.3), we get

$$-\frac{3P}{c^2}R^2\dot{R} = R\frac{d}{dt}(\rho R^2) + \rho R^2\frac{dR}{dt}.$$

This can be rewritten in the form (7.1.5).

Quite generally T^{00} represents the energy per unit volume or energy density (here $T^{00} = \rho c^2$). In the ultrarelativistic case of a gas of photons, or more generally of zero-rest-mass particles, each particle of energy $h\nu$ has momentum $h\nu/c$. Gases of this type are studied in thermodynamics; the energy density is related to the pressure by $\rho c^2 = 3P$.

We can study equation (7.1.5) in the ultrarelativistic case $\rho c^2 \approx 3P$, and the nonrelativistic case $P \ll \rho c^2$. In these cases the density ρ varies as R^{-4} and R^{-3}.

In the case of negligible pressure, equation (7.1.5) has the solution $\rho R^3 =$ constant, i.e.

$$\rho = \frac{\rho_0}{S^3} \quad \text{with} \quad S = \frac{R(t)}{R_0} \tag{7.1.6}$$

where R_0 and ρ_0 are the current values of $R(t)$ and $\rho(t)$.

Replacing ρ by this expression in equation (7.1.4), we get the Friedmann equation:

$$♦ \qquad \frac{1}{2}\dot{S}^2 - \frac{4\pi G\rho_0}{3}\frac{1}{S} = -\frac{kc^2}{R_0^2} = K = \text{constant}. \tag{7.1.7}$$

This equation may be discussed as in the Newtonian case (cf. II § 4.5). We note the correspondence between the relativistic and Newtonian models.

The reader will note that the spherical model corresponds to the Newtonian model in which the expansion phase is followed by a contraction phase leading to the *big crunch*.

The time between the *big bang* and the *big crunch* is $\dfrac{1}{H_0}\dfrac{2\pi q_0}{(2q_0 - 1)^{3/2}}$.

7.2 The Schwarzschild solution

To study the tests of metric theories in the Solar System (II chapter 3) we assumed that the gravitational field of the Sun is static and spherically symmetric. The Robertson expansions introduced two coefficients, β and γ; observation is compatible with $\beta = \gamma = 1$. We shall show here that the results are precisely those predicted by general relativity.

We consider the form (3.1.2) of the line element describing a gravitational field which is spherically symmetric but not necessarily static:

$$\blacklozenge \qquad ds^2 = e^\nu c^2 dt^2 - e^\lambda dr^2 - r^2 \sin^2\theta d\phi^2 - r^2 d\theta^2 \qquad (7.2.1)$$

where ν and λ are functions of r and t.

Using the spherical symmetry, we leave it as an exercise to show that the only non-zero components of the Ricci tensor associated with this metric are R_{00}, $R_{0r} = R_{r0}$, R_{rr}, and $R_{\phi\phi} = \sin^2\theta \, R_{\theta\theta}$.

7.2.1 Birkhoff's theorem

Direct calculation gives $R_{0r} = \dfrac{1}{rc}\dfrac{d\lambda}{dt}$.

The vacuum Einstein equations are $R_{\alpha\beta} = 0$; they therefore imply that λ is time-independent.

Thus we get

$$e^{\lambda-\nu} R_{00} \equiv \left(\frac{\nu''}{2} + \frac{\nu'^2}{4} + \frac{\nu'}{r} - \frac{\lambda'\nu'}{4}\right) = 0$$

$$R_{rr} \equiv -\left(\frac{\nu''}{2} + \frac{\nu'^2}{4} - \frac{\lambda'}{r} - \frac{\lambda'\nu'}{4}\right) = 0$$

$$R_{\theta\theta} \equiv -e^{-\lambda} + \frac{r}{2}(\lambda' - \nu')e^{-\lambda} + 1 \equiv -(re^{-\lambda})' + 1 - \frac{r}{2}e^{-\lambda}(\nu' + \lambda') = 0$$

with the notation $(\ )' = d(\ )/dr$.

Adding the first pair of equations, we find $\nu' + \lambda' = 0$. Using this relation, the last equation is easily integrated. We thus get

$$e^\lambda = \left(1 - \frac{a}{r}\right)^{-1} \quad \text{and} \quad e^\nu = \left(1 - \frac{a}{r}\right) e^{f(t)}$$

where $f(t)$ is an arbitrary function and a an arbitrary constant.

The functions e^ν and e^λ, to be solutions of the Einstein vacuum equations, must have this form. Direct verification shows that indeed these are solutions. However, the function $f(t)$ has no significance as it can be absorbed into the time coordinate by a change of the latter. Setting $f(t) = 1$ we get the general solution in the form

$$ds^2 = \left(1 - \frac{a}{r}\right) c^2 dt^2 - \frac{dr^2}{1 - \dfrac{a}{r}} - r^2 \sin^2\theta \, d\phi^2 - r^2 \, d\theta^2. \qquad (7.2.2)$$

·We see that a spherically symmetric solution of the Einstein vacuum equations is necessarily static. This is **Birkhoff's theorem**. The solution, called the Schwarzschild solution, thus has the form (7.2.2).

Let us assume that the gravitational field is created by a central body: our calculations apply to the region outside this body. The central object may undergo violent motions, such as a supernova explosion for example. However, if spherical symmetry continues to hold the metric remains unchanged as Schwarzschild. This is the physical meaning of Birkhoff's theorem.

> As an exercise, repeat in detail the calculations outlined here. Show in particular that the coordinate change defined by $dt' = e^{f(t)/2} dt$ implies $f(t) = 1$, up to notation.
> Show that the vacuum metric inside a spherical shell is Minkowski. To do this, use the continuity of $g_{\alpha\beta,\sigma}$ at the origin.

7.2.2 The Schwarzschild solution

We can make the change of variable

$$r = \bar{r}\left(1 + \frac{a}{4\bar{r}}\right)^2. \tag{7.2.3}$$

The line element (7.2.2) is then spatially isotropic:

$$ds^2 = \left(\frac{1 - \dfrac{a}{4\bar{r}}}{1 + \dfrac{a}{4\bar{r}}}\right)^2 c^2 dt^2 - \left(1 + \frac{a}{4\bar{r}}\right)^4 (d\bar{r}^2 + \bar{r}^2 \sin^2\theta \, d\phi^2 + \bar{r}^2 d\theta^2). \tag{7.2.4}$$

This expression can be compared with (3.1.3).

To interpret the coefficient a we consider a region very far from the origin, $r \gg a$. Expanding (7.2.4)

$$ds^2 = \left(1 - \frac{a}{\bar{r}} + \frac{1}{2}\left(\frac{a}{\bar{r}}\right)^2 + \ldots\right) c^2 dt^2$$
$$- \left(1 + \frac{a}{\bar{r}} + \ldots\right)(d\bar{r}^2 + \bar{r}^2 \sin^2\theta \, d\phi^2 + \bar{r}^2 d\phi^2). \tag{7.2.5}$$

To get agreement between general relativity and Newtonian theory we set $a = 2GM/c^2$, where M is the Newtonian mass of the central body producing the gravitational field (cf. II 3.3.2). The expansion (7.2.5) is then identical to that of Robertson (II 3.4.1) with $\alpha = 1$, $\beta = 1$ and $\gamma = 1$. These numerical values are exactly those observed, providing a positive test for general relativity.

> Tests of general relativity are not confined to tests of the Schwarzschild solution and measurements of β and γ. If spherical symmetry is abandoned the PPN expansions lead to the appearance of more than ten parameters whose numerical values are known in general relativity. Thus many post-Newtonian tests are possible; they are designed to measure the PPN parameters or combinations of them. For example laser telemetry gives the Earth–Moon distance to about one centimetre. The motion of the Moon may thus be compared with theoretical predictions taking into account post-Newtonian effects (and of course comparable or larger Newtonian effects). The PPN parameters can thus be found numerically. By 1996 none of the tests showed any disagreement with general relativity.

Given the coefficient a determined as above, the Schwarzschild solution (7.2.2) has the form

$$\blacklozenge \qquad ds^2 = \left(1 - \frac{2GM}{rc^2}\right)c^2 dt^2 - \frac{dr^2}{1 - \dfrac{2GM}{rc^2}} - r^2 \sin^2\theta \, d\phi^2 - r^2 d\theta^2. \quad (7.2.6)$$

The coordinates in this version of the metric are called Schwarzschild coordinates.

7.2.3 The Schwarzschild horizon

The Schwarzschild metric has a singularity at $r = R_S \equiv 2GM/c^2$; we recall that this quantity, called the Schwarzschild radius, was introduced in the first chapter in a Newtonian context.

Space-time is separated into two regions: $r < R_S$ "inside" the singularity and $r > R_S$ "outside" it.

For the Sun, $R_S \approx 2.9$ km. At the surface of the Sun the gravitational field is weak and the coordinate r can be interpreted as the radial coordinate in Newtonian theory; the Sun's surface corresponds to the value $r = R_0 \approx 7 \times 10^5$ km $> R_S$. This situation is very common in astrophysical objects, even for very dense ones. Thus a neutron star with a mass similar to the Sun has a radius R_0 of order ten kilometres, larger than its Schwarzschild radius. In these conditions Einstein's vacuum equations are valid only for $r > R_0 > R_S$; the Schwarzschild singularity therefore does not appear. There is no singularity of this type for $r < R_0$ as here the Einstein equations must be solved in the presence of matter.

It is possible that the Schwarzschild metric describes very compact astrophysical objects known as **black holes**, in which the singularity appears in vacuum.

Let us consider such an object.

We have already established the relation between the period T_s of a light signal emitted at $r = r_s$ and the period T_0, measured at $r = r_0$ (II § 3.3.2). In the Schwarzschild metric with $r_s > R_S$ and $r_0 > R_S$, we have

$$\frac{T_s}{T_0} = \frac{\sqrt{1 - \dfrac{R_S}{r_s}}}{\sqrt{1 - \dfrac{R_S}{r_0}}}. \quad (7.2.7)$$

The period T_s is fixed by the nature of the emitting source. For a source near the singularity, $r_s \approx R_S$, the observed period T_0 of the light signals is very long: the photon energy tends to zero as the source tends to the singularity. For all $r_0 > R_S$, observation of the central object is thus impossible. The situation is different from that encountered in studying Laplace's black object (II § 1.2).

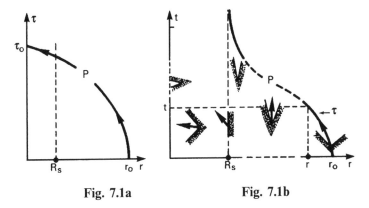

Fig. 7.1a　　　　　　**Fig. 7.1b**

If we assume that both regions, $r < R_S$ and $r > R_S$, are physically reasonable, we would like to know if they can communicate with each other.

Consider a test particle P in free fall, released from rest at the point $r = r_0 > R_S$. Its trajectory is a radial geodesic, parametrised by its proper time τ. As an exercise the reader can show that the equations of motion have the form

$$\frac{1}{2}\left(\frac{dr}{d\tau}\right)^2 - \frac{GM}{r} = -\frac{GM}{r_0} \tag{7.2.8}$$

with initial conditions $r = r_0 > R_S$ and $dr/d\tau = 0$ at the conventional value $\tau = 0$.

Equation (7.2.8) can be integrated by setting $r = r_0 \cos^2 \Psi$, where Ψ is a function of τ.

The origin $r = 0$ is reached in finite proper time τ_0, as is the Schwarzschild singularity, which is reached first (cf. fig. 7.1a).

Thus it is possible to communicate from the exterior to the interior. The point $r = R_S$ is in no way special for the motion considered; this suggests that the Schwarzschild singularity is not an essential singularity of space-time, but a singularity of the coordinate system. Moreover, study of the motion of P in the plane $\{r, t\}$ shows that the Schwarzschild singularity is reached in an infinite coordinate time t (cf. fig. 7.1b), although the proper time of P is finite. Schwarzschild coordinates are not well adapted to describe the crossing of the surface $r = R_S$, and other coordinate systems are generally used.

Consider the Euclidean plane $\{x, y\}$ with line element $d\ell^2 = dx^2 + dy^2$, and make the coordinate change

$$x = (1 - X) / \sqrt{|1 - X|},$$
$$y = Y.$$

A singularity now appears in the expression for the line element, as a result of the singularity of the coordinate system $\{X, Y\}$ at $X = 1$.

We note that (7.2.8) is a Friedmann equation.

Show that $\tau_0 = \dfrac{\pi}{2c} r_0 \sqrt{\dfrac{r_0}{R_S}}$.

For $r < R_S$, the metric coefficient g_{rr} is positive, while g_{00} is negative. This means that r is a time coordinate and t is a spatial coordinate (cf. II § 3.1). The arrow of time, i.e. the direction of changes in r corresponding to the ageing of observers for $r < R_S$, is chosen by continuity for the motion of P studied earlier. The singularity at $r = 0$ is thus temporal. This is an essential singularity as it persists in any coordinate system: it is a singularity of space-time itself.

In fig. 7.1b we have represented the light cones at various points of the plane $\{r, t\}$, and the possible trajectories for particles of non-zero rest mass. We see that a particle can never pass from the inside to the outside of the Schwarzschild singularity. For an external observer, the surface $r = R_S$ thus limits the region from which information can flow; the light cones are tangent to this surface at every point. This surface is called the **Schwarzschild horizon**.

Chapter 8

General relativity
in the weak field approximation

We assume that space-time may be covered by a quasi-Minkowski coordinate system (cf. II § 2.4.2). The line element may then be expanded as

$$\blacklozenge \quad \begin{cases} ds^2 = g_{\alpha\beta}dx^\alpha dx^\beta, \text{ with } g_{\alpha\beta} = \eta_{\alpha\beta} + h_{\alpha\beta} \\ \eta_{\alpha\beta} = \text{diag}(1,-1,-1,-1) \text{ and } |h_{\alpha\beta}| \text{ of the order of } \varepsilon \ll 1. \end{cases} \quad (8.0.1)$$

As usual Greek indices take values 0, 1, 2, 3, and Latin indices the values 1, 2, 3.

The coordinate systems we use are defined operatively by an experimental procedure associated with a procedure for calculating the coordinates of events. Both procedures can be specified; however, it is unnecessary to give them here.

> The measuring apparatus must be specified, and the method of using it and the calculation routines must be defined.

We shall study Einstein's equations when terms of order ε^2 are negligible. In the solar system for example, ε is less than 5×10^{-6} in Copernican coordinates; we are thus neglecting terms of order 10^{-11} by comparison with unity.

8.1 Conventions and formulae

Here we shall lower and raise indices using $\eta_{\alpha\beta}$ and its inverse $\eta^{\alpha\beta}$; for example

$$A^\alpha{}_\beta = \eta_{\beta\sigma}A^{\alpha\sigma} = \eta^{\alpha\sigma}A_{\sigma\beta}.$$

If $A^\alpha{}_\beta$ is a first-order quantity, the convention we use is equivalent up to the second order to the usual one in which we use the metric tensor $g_{\alpha\beta}$ or its inverse $g^{\alpha\beta}$ for lowering and raising indices.

In table 8.1 we specify the notation used and give expressions to the first order in ε for the various quantities used in the following. The d'Alembert operator is usually written "$\square$":

$$\square(\ ) = (\ )^{,\alpha}{}_\alpha = \frac{\partial^2(\ )}{c^2\partial t^2} - \Delta(\ ).$$

Table 8.1

Notation
  Coordinates:

$$(ct, x, y, z) = (x^0, x^1, x^2, x^3) = \{x^\alpha\} = x$$
$$(x^1, x^2, x^3) = \{x^k\} = (x, y, z) = \vec{r} = (r^k)$$
$$\|\vec{r}\| = r = \sqrt{x^2 + y^2 + z^2}, \quad \vec{r} \cdot \vec{r}' = x \cdot x' + y \cdot y' + z \cdot z'$$

  Partial derivatives and covariant derivatives:

$$\frac{\partial (\ )}{\partial x^\alpha} = \partial_\alpha (\ ) = (\ )_{,\alpha} \quad \text{and} \quad \nabla_\alpha (\ ) = (\ )_{;\alpha}$$

  Perturbations of the metric:

$$g_{\alpha\beta} = \eta_{\alpha\beta} + h_{\alpha\beta}, \quad g^{\alpha\beta} = \eta^{\alpha\beta} - h^{\alpha\beta}, \quad h_\alpha = h^\sigma{}_{\alpha,\sigma}, \quad h = h^\sigma{}_\sigma$$

Connection

$$\Gamma^\alpha{}_{\beta\sigma} = \{^\alpha{}_{\beta\sigma}\} = \frac{1}{2}(h^\alpha{}_{\beta,\sigma} + h^\alpha{}_{\sigma,\beta} - h_{\beta\sigma}{}^{,\alpha})$$

Riemann tensor, or curvature tensor

$$R_{\alpha\beta\mu\sigma} = \frac{1}{2}(h_{\alpha\mu,\beta\sigma} + h_{\beta\sigma,\alpha\mu} - h_{\beta\mu,\alpha\sigma} - h_{\alpha\sigma,\beta\mu})$$

Ricci tensor

$$R_{\beta\mu} \equiv R^\tau{}_{\beta\mu\tau} = \frac{1}{2}(h_{\mu,\beta} + h_{\beta,\mu} - h_{,\beta\mu} - \Box h_{\beta\mu})$$

Curvature scalar

$$R = R^\tau{}_\tau = h^\tau{}_{,\tau} - \Box h$$

The reader should verify the expressions of table 8.1, particularly the relation $g^{\alpha\beta} = \eta^{\alpha\beta} - h^{\alpha\beta}$.

8.2 Gauge invariance and the Einstein equations

Given a coordinate system, consider two space-times respectively provided with metrics $g_{\alpha\beta} = \eta_{\alpha\beta} + h_{\alpha\beta}$ and $g_{\alpha\beta} = \eta_{\alpha\beta} + \phi_{\alpha\beta}$, such that

$$\phi_{\alpha\beta} = h_{\alpha\beta} + \varepsilon_{\alpha,\beta} + \varepsilon_{\beta,\alpha} \tag{8.2.1}$$

where $\{\varepsilon^\mu(x)\}$ is a set of four arbitrary functions of $x = \{x^\alpha\}$, of first order in ε.

To the first order in ε, we can check the equality of the two Riemann tensors corresponding to each of the metrics (cf. table 8.1):

$$R_{\alpha\beta\sigma\mu}[h(x)] = R_{\alpha\beta\sigma\mu}[\phi(x)]. \qquad (8.2.2)$$

Thus there exists a family of metrics leading to the same expressions for the components of the Riemann tensor and thus the Ricci and Einstein tensors. The transformations (8.2.1) taking us from $h_{\alpha\beta}$ to $\phi_{\alpha\beta}$ are called **gauge transformations**. For tensors which are functions of the metric and its derivatives only, we call a gauge transformation the transformation given by substituting $\phi_{\alpha\beta}$ for $h_{\alpha\beta}$.

> The Riemann tensor describes the intrinsic properties of space-time, while the metric is more directly linked to the way in which the space-time is specified. For example in special relativity the Riemann tensor vanishes, while the metric components can take very different forms depending on the coordinate system used: Minkowski coordinates, coordinates comoving with an accelerated reference frame, spherical coordinates, etc. Thus many metrics may describe the same space-time. This property is analogous to gauge invariance in Maxwell's theory of electromagnetism.
>
> In Maxwell's theory the first of equations (I 6.2.1) allows us to write the electromagnetic field as $F_{\alpha\beta} = A_{\alpha,\beta} - A_{\beta,\alpha}$, where the A_α, called "four-potentials", are the components of a four-vector. We can check that the field is unchanged by the gauge transformation $A'_\alpha = A_\alpha + F_{,\alpha}$ where F is an arbitrary function of $\{x^\mu\}$. In general relativity, $R_{\alpha\beta\sigma\mu}$ represents "the gravitational field" and $g_{\alpha\beta}$ the "gravitational potentials". To first order, the gravitational field is unchanged if $h_{\alpha\beta}$, and thus $g_{\alpha\beta}$, are subject to a gauge transformation of the type (8.2.1).

Consider Einstein's equations in the form (6.4.6):

$$R_{\alpha\beta} = \chi \left(T_{\alpha\beta} - \frac{1}{2} g_{\alpha\beta} T \right) \qquad (8.2.3)$$

with $\chi = \dfrac{8\pi G}{c^4}$, where G is the Newtonian constant of gravitation.

We can imagine a theory in which $\chi = \varepsilon\chi_0$ where χ_0 has a given numerical value, with ε an arbitrary dimensionless parameter. The solutions of the Einstein equations are then functions of ε.

If ε were zero all gravitational interaction would disappear and space-time would be Minkowskian: we can assume that Minkowski coordinates were chosen such that $g_{\alpha\beta} = \eta_{\alpha\beta}$: to order zero $h_{\alpha\beta}$ is thus zero. Taking account of corrections to first order in ε, $h_{\alpha\beta}$ is found by solving the Einstein equations (8.2.3) in which $R_{\alpha\beta}$ is replaced by its expression in table 8.1.

From this point of view, χ is a first-order term in ε. In the Einstein equations (8.2.3), we can thus use the zero-order expression for $T_{\alpha\beta}$ and T on the right-hand side. We thus get the **linearised Einstein equations**.

To solve these equations we assume that the tensor $T_{\alpha\beta}$ is known to zero order. Writing the linearised Einstein equations in the form (6.4.5), it is easy to see that $T_{\alpha\beta}$ cannot be chosen arbitrarily, as to first order in ε we have

$S^{\alpha\beta}{}_{,\beta} = S^{\alpha\beta}{}_{;\beta} \equiv 0$. Thus we require the *integrability condition*

$$\partial_\beta T^{\alpha\beta} \equiv T^{\alpha\beta}{}_{,\beta} = 0. \tag{8.2.4}$$

These equations are just the "equations of motion" obeyed by the tensor $T^{\alpha\beta}$ in special relativity in a Minkowski system (cf. I chapter 6). The expressions for and interpretations of $T_{\alpha\beta}$ and $T^{\alpha\beta}$ to zero order are thus those of special relativity in a Minkowski coordinate system. We assume here that these quantities are known.

Assume that we have found a solution $h_{\alpha\beta}(x)$ of the linearised equations. The relation (8.2.2) tells us that $\phi_{\alpha\beta} = h_{\alpha\beta} + \varepsilon_{\alpha,\beta} + \varepsilon_{\beta,\alpha}$ is also a solution of the equations for the same source term $\chi T_{\alpha\beta}(x)$. We may use this property to look for a solution $\phi_{\alpha\beta}$ satisfying the following supplementary condition, known as the **harmonic condition** or **isothermal condition**. The gauge used is sometimes called the "de Donder gauge":

$$\blacklozenge \qquad\qquad \phi_\alpha - \frac{1}{2}\phi_{,\alpha} \equiv \phi^\sigma{}_{\alpha,\sigma} - \frac{1}{2}\phi^\sigma{}_{\sigma,\alpha} = 0. \tag{8.2.5}$$

The existence of $\phi_{\alpha\beta}$ can be shown as follows. Given a solution $h_{\alpha\beta}$ of the linearised Einstein equations, we set $\phi_{\alpha\beta} = h_{\alpha\beta} + \varepsilon_{\alpha,\beta} + \varepsilon_{\beta,\alpha}$ with $\Box\varepsilon_\alpha = -h_\alpha + h_{,\alpha}/2$. The existence of ε_α follows from general properties of the wave equation. It is then easy to check that the supplementary conditions (8.2.5) are satisfied by $\phi_{\alpha\beta}$. Thus, if the linearised Einstein equations have a solution, gauge invariance of the Ricci tensor guarantees the existence of a solution satisfying (8.2.5).

The linearised Einstein equations are then

$$\blacklozenge \qquad\qquad \Box\phi_{\alpha\beta} = -2\chi\left(T_{\alpha\beta} - \frac{1}{2}\eta_{\alpha\beta}\, T\right) \equiv -2\chi\Theta_{\alpha\beta}. \tag{8.2.6}$$

8.3 Solution of the linearised Einstein equations

If several sources of the gravitational field are present, in different parts of the Universe, for example, the solution of the linearised equations is the sum of the solutions found considering each of the sources independently. Here we consider one of the sources localised in the neighbourhood of the origin of coordinates, i.e. such that $T^{\alpha\beta} = 0$ for $r > a$. We will find the solution of the linearised Einstein equations for the case of slow motion, far from the sources, for $r \gg a$.

The source could also be localised near a point of arbitrary spatial coordinates b^k. We recover the present case by a coordinate change $x'^k = x^k - b^k$.

8.3.1 General form of solution

In the contravariant form the equations (8.2.6) can be written

$$\Box \phi^{\alpha\beta} = -2\chi\Theta^{\alpha\beta}. \tag{8.3.1}$$

The quantity $\phi^{\alpha\beta}$ is first order in ε. To find its significance, we can regard the coordinates as Minkowski coordinates in special relativity, which is the theory of order zero in ε, and regard $\phi^{\alpha\beta}$ as a field defined in Minkowski space. With this interpretation, equation (8.3.1) characterises a phenomenon propagating with speed c. Moreover, among the solutions of equation (8.3.1), we must retain only those for which the Riemann tensor, describing the properties of the space-time, satisfies the principle of causality. This suggests that we should retain the retarded solution:

$$\blacklozenge \qquad \phi^{\alpha\beta}(\vec{r},t) = -\frac{\chi}{2\pi} \iiint \frac{1}{R} \Theta^{\alpha\beta}\left(\vec{r}\,',t - \tfrac{R}{c}\right) d^3r', \tag{8.3.2}$$

with $R = \sqrt{(\vec{r}\,' - \vec{r})^2}$. The integration is over all space; in such cases we adopt the convention of not writing the limits of the integrals explicitly. In practice the integration is only over the region $r' < a$, outside of which $T^{\alpha\beta}$ vanishes as $\Theta^{\alpha\beta}$ vanishes with $T^{\alpha\beta}$. The volume element $dx'dy'dz'$ is written d^3r'.

Using the conservation relation (8.2.4), we can verify directly that the solution retained satisfies the gauge condition (8.2.5).

We omit the indices for brevity. The solutions of equation (8.3.1) appear as a linear combination of the three quantities $\phi^0, \phi^{(+)}$ and $\phi^{(-)}$.

The quantity ϕ^0 is the general solution of the linearised equations with zero right-hand side. The solution $\phi^0 = f(t - z/c)$, for example, describes free **gravitational** waves not corresponding to any source. The quantity ϕ^0 can also describe an inertial field; in this case the corresponding Riemann tensor vanishes. A coordinate change allows us to absorb this quantity; we can show this using the results of Appendix B concerning the Minkowski character of spaces in which the Riemann tensor vanishes. We do not retain these terms here.

The quantities $\phi^{(-)}$ and $\phi^{(+)}$ are expressed as

$$\phi^{(-)}(\vec{r},t) = -\frac{\chi}{2\pi} \iiint \frac{1}{R} \Theta\left(\vec{r}\,',t - \tfrac{R}{c}\right) d^3r'$$

$$\phi^{(+)}(\vec{r},t) = -\frac{\chi}{2\pi} \iiint \frac{1}{R} \Theta\left(\vec{r}\,',t + \tfrac{R}{c}\right) d^3r'.$$

The solutions $\phi^{(-)}$ and $\phi^{(+)}$ decrease as r^{-1} as r tends to infinity. Far from the sources the coordinate system is thus a Minkowski system in special relativity.

The state of the source, $\Theta(\vec{r}\,',t + R/c)$, on the future light cone of the point $\{t,\vec{r}\}$ determines $\phi^{(+)}$ at this point. For physical reasons we postulate that the "future" cannot influence the "past": this is the principle of causality. To satisfy this principle we discard the solution $\phi^{(+)}$ and retain only the solution (8.3.2), which gives a Riemann tensor satisfying the principle of causality. Gauge changes modify the metric, but not this property.

Consider a region very distant from the source, $r \gg a$ (fig. 8.1). We replace R by its expansion, neglecting, in expression (8.3.2) for $\phi^{\alpha\beta}$, terms decreasing

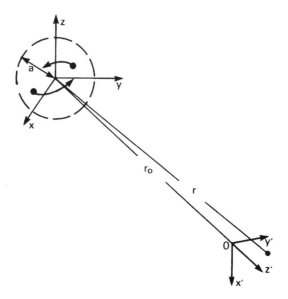

Fig. 8.1

as $(a/r)^n$, with $n > 1$, as r tends to infinity. We thus obtain

$$\phi^{\alpha\beta}(\vec{r}, t) = -\frac{\chi}{2\pi}\frac{1}{r} \iiint \Theta^{\alpha\beta}\left(\vec{r}', t - \tfrac{R}{c}\right) d^3r', \quad \text{with } R = r - \frac{\vec{r}\cdot\vec{r}'}{r}. \qquad (8.3.3)$$

The terms decreasing like r^{-1} can be omitted from the expression for R, as expansion of $\Theta^{\alpha\beta}$ for r large enough shows that these terms bring to $\phi^{\alpha\beta}$ a contribution of the type we have neglected here.

8.3.2 Description of the source

To describe the source $T^{\alpha\beta}$ of the gravitational field, we define the following **multipoles**:

$$\begin{cases} M = \dfrac{1}{c^2} \iiint T^{00}(\vec{r}, t)d^3r & : \text{ monopole} \\[2mm] M^k = \dfrac{1}{c^2} \iiint T^{00}r^k d^3r & : \text{ dipole} \\[2mm] M^{jk} = \dfrac{1}{c^2} \iiint T^{00}r^j r^k d^3r & : \text{ quadrupole.} \end{cases} \qquad (8.3.4)$$

We note that the various multipoles are functions of t alone. The conservation relation (8.2.4) allows us to prove the following:

$$\begin{cases} \dot{M} = 0, \quad \dot{M}^k = \dfrac{1}{c} \iiint T^{0k}(\vec{r}, t)d^3r \equiv P^k, \quad \dot{P}^k = \ddot{M}^k = 0 \\[3mm] \dfrac{1}{2}\ddot{M}^{kj} = \iiint T^{kj}(\vec{r}, t)d^3r = \dfrac{1}{c}\dfrac{d}{dt} \iiint r^k T^{0j}(\vec{r}, t)d^3r \end{cases} \qquad (8.3.5)$$

where we have used the notation $(\dot{\;}) = c\partial_0(\;) = \partial_t(\;)$.

In chapter 6 of part one we interpreted the quantity $T^{0\alpha}/c$ as a four-momentum density; Mc^2 thus represents the total energy of the source and M its inertial mass, while P^k represents the spatial components of its momentum. The relations $\dot{M} = 0$ and $\dot{P}^k = 0$ thus express the conservation in time of the energy and momentum of the matter distribution which is the source of the gravitational field.

The relation (8.2.4) can be written explicitly as

$$\frac{\partial T^{0\alpha}}{c\partial t} \equiv T^{0\alpha}{}_{,0} = -T^{j\alpha}{}_{,j} = -T^{\alpha j}{}_{,j}.$$

Using the definition of M we thus get

$$\dot{M} \equiv \frac{1}{c}\iiint T^{00}{}_{,0}d^3r = -\frac{1}{c}\iiint T^{0k}{}_{,k}d^3r.$$

This expression is the integral of a divergence. Using Gauss's theorem, we can express it as an integral of T^{0k} over an arbitrary surface *enclosing* the source. On such a surface T^{0k} vanishes; we thus obtain the first of the relations (8.3.5).

The definition of M^k and the conservation equation (8.2.4) lead to the relations

$$\dot{M}^k = \frac{1}{c}\iiint T^{00}{}_{,0}r^k d^3r = -\frac{1}{c}\iiint T^{0j}{}_{,j}r^k d^3r = \frac{1}{c}\iiint [T^{0k} - (T^{0j}r^k)_{,j}]d^3r.$$

As before the use of Gauss's theorem gives

$$\dot{M}^k = \frac{1}{c}\iiint T^{0k}d^3r \Rightarrow \ddot{M}^k = \iiint T^{0k}{}_{,0}d^3r = -\iiint T^{jk}{}_{,j}d^3r = 0.$$

The proof of the last relations of (8.3.5) follows the same method:

$$c\dot{M}^{jk} = \iiint T^{00}{}_{,0}r^j r^k d^3r = -\iiint T^{0m}{}_{,m}r^j r^k d^3r$$

$$= -\iiint [(T^{0m}r^j r^k)_{,m} - (T^{0j}r^k + T^{0k}r^j)]d^3r$$

$$= \iiint (T^{0j}r^k + T^{0k}r^j)d^3r.$$

After differentiation with respect to time

$$\ddot{M}^{jk} = \iiint (T^{0j}{}_{,0}r^k + T^{0k}{}_{,0}r^j)d^3r = -\iiint (T^{mj}{}_{,m}r^k + T^{mk}{}_{,m}r^j)d^3r$$

$$= -\iiint [(T^{mj}r^k + T^{mk}r^j)_{,m} - 2T^{jk}]d^3r = \iiint 2T^{jk}d^3r.$$

The reader should verify the last of the relations (8.3.5).

We assume now that the source moves slowly with respect to the coordinate system we have introduced. This hypothesis can be used in expanding the quantity $T^{\alpha\beta}(\vec{r}\,',t-R/c)$ which appears in (8.3.2), in powers of $1/c$.

$$T^{\alpha\beta}\left(\vec{r}\,',t-\frac{R}{c}\right) = T^{\alpha\beta}\left(\vec{r}\,',t-\frac{r}{c}\right) + \frac{\vec{r}\cdot\vec{r}\,'}{cr}\frac{\partial}{\partial t}T^{\alpha\beta}\left(\vec{r}\,',t-\frac{r}{c}\right)$$
$$+\frac{1}{2}\left(\frac{\vec{r}\cdot\vec{r}\,'}{cr}\right)^2\frac{\partial^2}{\partial t^2}T^{\alpha\beta}\left(\vec{r}\,',t-\frac{r}{c}\right)+\dots \quad (8.3.6)$$

To use (8.3.6) coherently, we must take account of the fact that the various components of the tensor $T^{\alpha\beta}$ are not of the same order in $1/c$: $T^{jk} \approx (1/c)T^{0k} \approx (1/c)^2 T^{00}$.

> Expanding $T^{\alpha\beta}$ in Fourier integrals, show that the approximation of slow motion implies that the frequencies ν characterising the motion are small; more precisely $\lambda \equiv c/\nu >> a$.

Limiting the expansions to the first order in ε we can regard the coordinate system as Minkowskian for the interpretation of $T^{\alpha\beta}$. Considering the case of slow motions, we can identify this system with a Galilean system of Newtonian theory. We shall use Newtonian language in what follows wherever possible. To interpret the effects of the perturbation $h_{\alpha\beta}$ we must however make use of the physical significance of the metric; to do this we must abandon the Newtonian interpretation.

8.3.3 Solution of the linearised Einstein equations

We use the expansion (8.3.6) of $T^{\alpha\beta}$ in powers of $1/c$ to express the term $\Theta^{\alpha\beta}$ on the right-hand side of the equation to be solved (8.3.1). Expansion of the solution (8.3.3) gives the expression for $\phi_{\alpha\beta}$. A suitable change of gauge, $h_{\alpha\beta} = \phi_{\alpha\beta} + \varepsilon_{\alpha,\beta} + \varepsilon_{\beta,\alpha}$, then gives the following particular solution:

$$\blacklozenge \quad \begin{cases} h_{00}(\vec{r},t) = -\dfrac{2G}{rc^2}\left(M-\dfrac{1}{c}n_k P^k\right) \quad h_{0k}(\vec{r},t) = -\dfrac{4G}{rc^3}P^k \\[3mm] h_{jk}(\vec{r},t) = -\dfrac{2G}{rc^2}\left(-\eta_{jk}\left(M-\dfrac{1}{c}n_m P^m\right)+\dfrac{1}{c^2}P_{jkmn}\ddot{\mathbb{I}}^{mn}\left(t-\dfrac{r}{c}\right)\right) \end{cases} \quad (8.3.7)$$

where we have used the definitions

$$\mathbb{I}^{mn} = M^{mn} - \frac{1}{3}\eta^{mn}M^k{}_k \quad P_{jkmn} = \Pi_{jm}\Pi_{kn} - \frac{1}{2}\Pi_{jk}\Pi_{mn} \quad (8.3.8)$$

with $\Pi_{jk} = \eta_{jk} + n_j n_k$ and $n_k = \eta_{kj}n^j = -n^k = -\dfrac{r^k}{r}$.

The general solution of the Einstein equations is then given by making an arbitrary gauge transformation on the solution $h_{\alpha\beta}$. This property expresses **the gauge invariance of linearised theory**; it also expresses the freedom in the metric if only Einstein's equations are solved.

To obtain expressions (8.3.7), we determine $\phi_{\alpha\beta}$:

$$\phi_{00} = -\frac{\chi}{4\pi r}\left(Mc^2 - n_m\dot{M}^m c + \frac{1}{2}n_m n_n\ddot{M}^{mn} - \frac{1}{2}\ddot{M}^m{}_m\right)$$

$$\phi_{0k} = -\frac{\chi}{2\pi r}\left(\dot{M}_k c - \frac{1}{2}n_m\ddot{M}^m{}_k\right)$$

$$\phi_{jk} = -\frac{\chi}{4\pi r}\left[-\eta_{jk}\left(Mc^2 - n_m\dot{M}^m c + \frac{1}{2}n_m n_n\ddot{M}^{mn} + \frac{1}{2}\ddot{M}^m{}_m\right) + \ddot{M}_{jk}\right].$$

The functions M^{jk} are evaluated for the argument $t - r/c$; M and $\dot{M}^k = P^k$ are constants; the values of $\phi_{\alpha\beta}$ are those at the point $\vec{r}$ and time t.

We make the gauge change $\mathbf{h}_{\alpha\beta} = \phi_{\alpha\beta} + \varepsilon_{\alpha,\beta} + \varepsilon_{\beta,\alpha}$, with the following values:

$$\varepsilon_0 = \frac{\chi c}{16\pi r}(n_m n_n\dot{M}^{mn} - \dot{M}^m{}_m),$$

$$\varepsilon_k = -\frac{\chi c}{4\pi r}\left(n_m\dot{M}^m{}_k + \frac{1}{4}n_k n_m n_n\dot{M}^{mn} - \frac{1}{4}n_k\dot{M}^m{}_m\right).$$

We thus get (8.3.7) for the approximations adopted. In performing the calculations it is useful first to verify the following properties:

$$\ddot{M}^m{}_m = 0, \quad \left(\frac{1}{r}\right)_{,k} = O(r^{-2}), \quad \frac{1}{r}n^k{}_{,j} = O(r^{-2}), \quad \left(F\left(t - \frac{r}{c}\right)\right)_{,k} = \frac{n_k}{c}\dot{F}\left(t - \frac{r}{c}\right)$$

for any function F. $O(r^{-2})$ denotes a quantity decreasing like r^{-2} as r tends to infinity, and which may therefore be neglected here.

8.4 Changes of coordinates

We begin by considering first order coordinate changes so as to interpret gauge changes. In this context we consider the problem of measurement, and we show how the results of a certain class of measurements called "gauge invariant measurements" fit with our incomplete knowledge of the metric in the given coordinate system.

> The gauge invariance of the theory has appeared as a mathematical property of the linearised Einstein equations. We shall interpret this property by considering two observers with neighbouring coordinate systems. For these observers the components of the Riemann tensor are represented by identical functions. The same holds for the source term in the Einstein equations, $\chi\Theta^{\alpha\beta}$. The two metrics will however differ, as their perturbations from the Minkowski values can be deduced from each other by a gauge transformation.

We complete the study of coordinate changes by considering Poincaré transformations.

8.4.1 First-order coordinate changes

We consider a first observer O_1, with a coordinate system SC_1. We consider a second observer O_2 whose coordinate system is SC_2 close to the first, either because the measuring apparatus is slightly changed, or because the procedure of calculating is slightly altered. A first-order coordinate change is characterised by the relations

$$x_2{}^\alpha = x_1{}^\alpha + \varepsilon^\alpha(x_1). \tag{8.4.1}$$

The functions $\varepsilon^\alpha(x)$ are assumed to be of the same order as ε. To first order in ε, these expressions can be inverted in the form

$$x_1{}^\alpha = x_2{}^\alpha - \varepsilon^\alpha(x_2). \tag{8.4.2}$$

Consider a field of tensors whose components in the two coordinate systems are $A_{1\alpha\beta\dots}(x_1)$ and $A_{2\alpha\beta\dots}(x_2)$. The formulae for coordinate changes give the first order relations:

$$A_{2\alpha\beta\dots}(x_2) = (A_{1\alpha\beta\dots} - \varepsilon_{\sigma,\alpha}A_1{}^\sigma{}_{\beta\dots} - \varepsilon_{\sigma,\beta}A_{1\alpha}{}^\sigma{}_{\dots} - \dots)_{(x_1)}$$

$$A_{2\alpha\beta\dots}(x_2) = (A_{1\alpha\beta\dots} - \varepsilon_\sigma A_{1\alpha\beta\dots,}{}^\sigma - \varepsilon_{\sigma,\alpha}A_1{}^\sigma{}_{\beta\dots} - \varepsilon_{\sigma,\beta}A_{1\alpha}{}^\sigma{}_{\dots} - \dots)_{(x_2)}$$

We note that first order tensors remain invariant to the second order:

$$A_{1\alpha\beta\dots}(x_1) = A_{2\alpha\beta\dots}(x_1) = A_{1\alpha\beta\dots}(x_2) = A_{2\alpha\beta\dots}(x_2).$$

This is true for example of the Riemann and Ricci tensors; it is also true for the source tensor of the gravitational field appearing on the right-hand side of the Einstein equations (8.3.1). In the latter case, $T_{\alpha\beta}$ and T are modified in the coordinate change by a first order term, whose contribution becomes negligible after multiplication by χ.

The formula for coordinate changes may also be applied to the metric coefficients; we get

$$\begin{cases} g_{1\alpha\beta} = \eta_{\alpha\beta} + h_{1\alpha\beta}, \ g_{2\alpha\beta} = \eta_{\alpha\beta} + h_{2\alpha\beta}, \text{ with} \\ h_{2\alpha\beta}(x_2) = h_{1\alpha\beta}(x_2) - \varepsilon_{\alpha,\beta}(x_2) - \varepsilon_{\beta,\alpha}(x_2). \end{cases} \tag{8.4.3}$$

The quasi-Minkowski form of the metric is preserved. The functions h_2 and h_1 correspond to each other under a gauge transformation up to second order. We note that $h_{\alpha\beta}$ does not transform as a tensor.

First order tensors which can be expressed as functions of $h_{\alpha\beta}$ and its derivatives are invariant under coordinate changes as they are first order tensors. They undergo a gauge transformation when we make a coordinate change. They are therefore invariant under this gauge transformation, which is arbitrary, given the arbitrary nature of the coordinate change considered. This is true of the Riemann tensor and the tensors obtained from it by contraction.

These mechanisms explain the gauge invariance of the theory. In fact, given a gauge transformation, this can be regarded as the change of $h_{\alpha\beta}$ in a coordinate change.

We can make coordinate changes characterised by functions of order ε' which differs from ε. The results are the same provided we neglect terms in ε^2, ε'^2 and $\varepsilon\varepsilon'$.

If we are only interested in the properties of the space-time itself, we can study these equally well in SC_1 or SC_2. The choice of gauge is thus only a question of convenience. This is not the case if we are interested in precise experiments which require the use of a precise coordinate system, set up by the observer. In such cases we have to know the metric in the coordinate system used to be able to predict the results up to first order.

Let us identify ourselves with an observer equipped with operationally defined coordinates SC_1. As an example, we consider an experiment in which a clock is maintained at rest in SC_1, which we may control experimentally. We propose to determine the proper time f_1 of this clock between two given events A and A'. f_1 depends on the time coordinates t_1 and t_1' of A and A', determined by the reference clocks of SC_1, as well as the position of the experiment determined by the coordinates x_1^k, which are held constant:

$$f_1 = t_1' - t_1 + \frac{1}{2} \int_{t_1}^{t_1'} h_{100}(t'', x_1^k) dt''.$$

To predict the result we have to know h_{100}, or more generally $h_{1\alpha\beta}$, if we wish to consider an arbitrary experiment.

The same experimental procedure relative to a second observer leads to $dx_2^k/dt_2 = 0$; this is not the procedure associated with the same experiment. We assume however that in the two experiments the events A and A' are the same and that the clocks meet at A and A'. The predictions of the two observers will differ (cf. the "twin paradox" in special relativity). The prediction of the second observer depends on $h_{2\alpha\beta}$. For a given procedure, the theoretical predictions and experimental results depend on the coordinate system in which the procedure is carried out. They are not invariant under changes of gauge.

We must distinguish clearly between the situation above and the description of the same experiment by two observers. In this case, in the example above there would be only one clock, at rest in SC_1 and possibly moving in SC_2. The experiment is described by different procedures by each of the observers, but the predictions remain the same.

In some cases however the results of measurements are independent of the gauge, and we can then tolerate some lack of knowledge about the metric associated with the coordinate system used. In such a situation we can use for $h_{\alpha\beta}$ the expression (8.3.7) for $\mathbf{h}_{\alpha\beta}$. This kind of measurement is called "gauge invariant".

Let Q be the result of the measurement considered. A theoretical description of the measurement gives an expression for Q which contains a number of numerical quantities defining the experimental procedure in the coordinate system considered. These quantities can be numbers or functions. In general the theoretical expression for Q also depends on the metric. We assume that Q is a first order quantity. If Q depends on the metric through a gauge-invariant tensor, the Riemann tensor for example, the measurement is gauge-invariant. We can thus assume for calculating Q that the gauge is that which gives the metric perturbation specified by the relations (8.3.7).

Consider two neighbouring quasi-Minkowski coordinate systems; the system SC_1 defines in an operational way the system SC_0 in which $h_{\alpha\beta} = \mathbf{h}_{\alpha\beta}$.

The numerical quantities defining the experimental procedures for the measurement considered are denoted symbolically by $\mathbf{A}_1$ and $\mathbf{A}_0$ in the two coordinate systems. The result of a measurement is a scalar, and we can calculate Q equally well in SC_1 or SC_0: the results are identical. In SC_1 we can use the procedure defined by $\mathbf{A}_1$, but we do not know the gauge. In SC_0, we know the metric but we do not know $\mathbf{A}_0$. However, $\mathbf{A}_0$ differs from $\mathbf{A}_1$ by a quantity of first order in ε. Assume that Q is a first-order quantity; to second order we may thus perform the calculation in SC_0 with $\mathbf{A}_0 = \mathbf{A}_1$ and $h_{\alpha\beta} = \mathbf{h}_{\alpha\beta}$. Hence we can assume the equality $h_{1\,\alpha\beta} = \mathbf{h}_{\alpha\beta}$. The measurement is gauge-invariant in this case.

In the general case we get

$$Q = Q[\mathbf{A}, g_{\alpha\beta}] = Q_0[\mathbf{A}, \eta_{\alpha\beta}] + Q_1[\mathbf{A}, h_{\alpha\beta}].$$

Q is a result of measurement, $Q[\mathbf{A}, g_{\alpha\beta}]$ is given by theory; it is a functional of the metric and the quantities $\mathbf{A}$ which define the measurement procedure in the quasi-Minkowski system used, $Q_0[\mathbf{A}, \eta_{\alpha\beta}]$ and $Q_1[\mathbf{A}, h_{\alpha\beta}]$ represent the principal term and gravitational perturbation of the first order. In reality, $\mathbf{A}$ is not perfectly known. The true value of $\mathbf{A}$ is determined to a first approximation by relations of the type $Q = Q_0[\mathbf{A}, \eta_{\alpha\beta}]$ by numerical fitting. In doing this we determine $\mathbf{A}$ by identifying the quasi-Minkowski system to a rigorously Minkowski system, and commit an error of first order in $\mathbf{A}$, leading to a second order error only in the expression for $Q_1[\mathbf{A}, h_{\alpha\beta}]$. The first order gravitational perturbation is thus found as if the measuring apparatus were situated in a Minkowski space-time and were sensitive to a field, $h_{\alpha\beta}$, defined on that space-time. In this case Q_1 and Q_0 are not separately scalars and the interpretation of the measurement does not overcome the ignorance of the gauge. We must then increase the measurements to determine not only $\mathbf{A}$ but also $h_{\alpha\beta}$ by numerical fitting. The interest in measurements such as $Q_0[\mathbf{A}, \eta_{\alpha\beta}] = 0$ is clear; in this case we need only determine $h_{\alpha\beta}$ up to a gauge transformation.

As an exercise, the reader should examine the experiments described in chapter 3 in this way.

8.4.2 Poincaré transformations

Consider the following change of coordinates:

$$x^\alpha = A^\alpha{}_\mu x'^\mu + a^\alpha \quad \text{with} \quad A^\alpha{}_\mu A^\beta{}_\sigma \eta_{\alpha\beta} = \eta_{\mu\sigma} \tag{8.4.4}$$

where $A^\alpha{}_\mu$ and a^α are constants.

A coordinate change of this type, which conserves the Minkowski form of the metric in special relativity, is called a *Poincaré transformation* (cf. I § 2.2). In the new coordinate system the metric is

$$g'_{\alpha\beta}(x') = \eta_{\alpha\beta} + h'_{\alpha\beta}(x') \quad \text{with} \quad h'_{\alpha\beta}(x') = A^\mu{}_\alpha A^\sigma{}_\beta h_{\mu\sigma}(x). \tag{8.4.5}$$

Thus, under a Poincaré transformation, a quasi-Minkowski metric retains its quasi-Minkowski form. The perturbation $h_{\alpha\beta}(x)$ transforms like a field of tensors in Minkowki space, where the coordinates $x = \{x^\alpha\}$ are interpreted as Minkowski coordinates.

Consider spatial rotations, for which η_{00}, η_{0k} and η_{kj} remain invariant. We can verify that h^{00}, h^{0k} and h^{jk} transform as scalar, vector and second-rank tensor respectively. This holds also for M, M^k, M^{kj} and $\maltese^{mn}$, while P_{jkmn} transforms as a tensor of rank four, and n^k as a vector. The vectors and tensors here are purely spatial quantities; whether they are covariant or contravariant is determined by the position of their Latin indices, lowered or raised.

In a given coordinate system SC_1, a suitable gauge transformation puts $h_{\alpha\beta}$ into the form (8.3.7). This implies that $\mathbf{h}_{\alpha\beta}$ is the expression for $h_{\alpha\beta}$ in a coordinate system SC'_1 close to SC_1. Under a spatial rotation SC'_1 becomes the coordinate system SC'_2. The perturbation remains form-invariant. This property is a consequence of the identical tensor nature of the two sides of the relations (8.3.7). We get the same result by solving the Einstein equations in the coordinate system SC_2 deduced from SC_1 by the rotation considered and making the gauge transformation leading to the form (8.3.7).

It is useful to distinguish between the expressions "change of coordinates" and "gauge transformation", even if the gauge transformation is interpreted as a first-order change of coordinates. In a "gauge transformation" the new perturbation is determined with respect to an underlying metric whose components remain unchanged functions of the coordinates. In a "change of coordinates" it is the metric components and the perturbation which transform independently. For first-order transformations these two operations, which are equivalent when we consider the metric, are very different when we express the perturbation of the metric.

8.5 Interpretation of $\mathbf{h}_{\alpha\beta}$

The expressions (8.3.7) for $\mathbf{h}_{\alpha\beta}$ contain three kinds of terms: monopole terms proportional to M, dipole terms proportional to P^k, and quadrupole terms proportional to $\ddot{\maltese}^{km}$. The monopole, dipole and quadrupole terms are respectively of relative order 1, $1/c$ and $1/c^2$.

Monopole terms

Neglecting multipole terms of order higher than unity, the expressions (8.3.7) reduce to the expansion to first order in r^{-1} of the Schwarzschild metric in isotropic coordinates (cf. II § 7.2.2). Far from the source, the source appears as a collection of matter with spherical symmetry to a first approximation. The coefficient M, already interpreted as the inertial mass of the source in the sense that it is a measure of its total energy (cf. § 8.3.2), must also be understood as its gravitational mass. Here we recover in a sense the Newtonian equivalence between inertial and gravitational mass.

Dipole terms

We have interpreted P^k as the momentum of the source of the gravitational field (cf. § 8.3.2). The corresponding dipole terms appear in the expression for $\mathbf{h}_{\alpha\beta}$ if the source has a global motion with respect to the coordinate system. These terms disappear if the coordinate system is chosen so that the centre of mass of the source is at rest in it. The dipole terms thus do not describe any intrinsic property of the space-time.

As in special relativity, the relation $T^{\alpha\beta}{}_{,\beta} = 0$ implies that $\{Mc, P^k\}$ are constant. The reader should show similarly that under a Poincaré transformation, $\{Mc, P^k\}$ transform as a four-vector which to zero order in ε is the energy–momentum of the matter distribution. A Poincaré transformation allows us to set $P^k = 0$. In this case the calculation of $h_{\alpha\beta}$ follows as before, with $P^k = 0$.

Quadrupole terms and gravitational waves

Consider the quadrupole terms in the expression for $\mathbf{h}_{\alpha\beta}$. Using the summation convention for repeated indices, these can be expressed as:

$$h^{(W)}{}_{jk}(\vec{r}, t) = -\frac{2G}{rc^4}\, P_{jkmn}\, \ddot{I}^{jmn}\left(t - \frac{r}{c}\right). \qquad (8.5.1)$$

In the coordinate system considered, the quadrupole terms are the only ones which vary in time; this feature gives them an "observable" character which picks them out in the expression (8.3.7).

The reader should check that I^{mn} is independent of the origin chosen for the calculation. Consider a source with spherical symmetry. Choosing the centre of symmetry as origin, show that I^{mn} is zero. The time-dependent terms in the expressions (8.3.7) vanish. The metric is time-independent in the corresponding coordinate system. We recover here Birkhoff's theorem.

The quantities $h^{(W)}{}_{jk}$ have a wavelike structure: they satisfy the propagation equation $\Box h^{(W)}{}_{jk} = 0$ and their amplitudes decrease as $1/r$ for large r. These quantities describe **gravitational radiation** or gravitational waves, which propagate in vacuo at the speed of light.

The reader should not confuse "gravitational waves" with "gravity waves". An example of the latter are water waves: the restoring force acting on the water is its weight, i.e. gravity.

We consider a region of space surrounding the point O, very far from the source. The size of this region is small enough that we can regard r as constant (fig. 8.1). We make a spatial rotation of the coordinate system described by the rotation matrix $R^k{}_j$ so that the coordinates of O become $\{0, 0, r_0\}$. A translation of the origin brings the centre of the coordinate system to O.

The time coordinate is unchanged: the new space coordinates can be expressed in terms of the old ones in the form:

$$x'^k = R^k{}_j x^j + a^k \quad \text{and} \quad \{a^k\} = \{0, 0, -r_0\}.$$

In this coordinate system, in the neighbourhood of O, the only non-zero components of the quadrupole term are:

$$h'^{(W)}{}_{11} = -h'^{(W)}{}_{22} = f_+\left(t - \frac{z'}{c}\right) \qquad h'^{(w)}{}_{12} = h'^{(w)}{}_{21} = f_\times\left(t - \frac{z'}{c}\right) \qquad (8.5.2)$$

$$f_+ = -\frac{G}{r_0 c^4}(\mathbf{\dddot{\mathtt{I}}}^{'11} - \mathbf{\dddot{\mathtt{I}}}^{'22})_{(t-\frac{z'}{c})} \quad f_\times = -\frac{2G}{r_0 c^4}(\mathbf{\dddot{\mathtt{I}}}^{'12})_{(t-\frac{z'}{c})} \tag{8.5.3}$$

with $\mathbf{\mathtt{I}}^{'jk} = R^j{}_m R^k{}_n \mathbf{\mathtt{I}}^{mn}$.

The gravitational wave locally has the form of a plane wave of speed c. Near O the other terms in $\mathbf{h}_{\alpha\beta}$ remain effectively constant, independent of time and position in the neighbourhood of O. Under these conditions, a suitable first-order coordinate change gives the metric form

$$g''_{\alpha\beta}(x'') = \eta_{\alpha\beta} + h''^{(W)}{}_{\alpha\beta}(x'') \quad \text{with}$$

$$h''^{(W)}{}_{\alpha\beta}(x'') = h'^{(W)}{}_{\alpha\beta}\left(t'' - \frac{z''}{c}\right) \quad \text{and} \quad z''_0 = 0. \tag{8.5.4}$$

To obtain (8.5.2) and (8.5.3), we use the formula (8.3.7) with $\{n^k\} = \{0, 0, 1\}$, as well as the relation $\mathbf{\mathtt{I}}^m{}_m = 0$ and the tensor properties of $\mathbf{\mathtt{I}}^{mn}$. As an exercise, the reader should find the coordinate change leading to the form (8.5.4).

Thus a gravitational wave in vacuo is locally determined by two independent functions f_+ and $f_\times$. We can also say that *a gravitational wave has two polarisations*. These results can be found using the vacuum Einstein equations near the point O without introducing the sources of the gravitational field. By considering plane gravitational waves which propagate along Oz, the reader should show as an exercise the existence of a harmonic coordinate system in the vicinity of O such that $h_{11} = -h_{22}$ and $h_{12} = h_{21}$ are the only non-zero components of $h_{\alpha\beta}$. Note that the gravitational field is not necessarily weak everywhere, but only near O.

We have presented weak field theory as the first order in an expansion in powers of an arbitrary parameter ε. The extension to higher orders is not without difficulty.

The approximations we have made are justified by current observations. However, these approximations are not always adequate. For example, the perihelion advance of satellites is a second-order effect in ε which is actually observed. Further, this theory is of no use for dealing with the most compact objects.

Chapter 9

Two experimental projects:
a satellite gyroscope and a gravitational wave detector

In the first part of this chapter, the study of a satellite gyroscope gives the chance to use the ideas introduced in this book in a practical context. Considerable prior knowledge is therefore required.

In the second part, on gravitational waves, the aim is different. The reader may assume the results of the previous chapter in order to follow this part.

General relativity belongs to the category of metric theories of gravity. Several observations of various types eliminate many of these theories, while general relativity currently survives unscathed: there is as yet no reason to doubt the validity of the theory in any domain where it has been tested.

Among the predictions of general relativity, two have been the subjects of significant development efforts for several decades, with the aim of verifying them or exploiting their consequences. These are the motion of a satellite-borne gyroscope in the Earth's gravitational field, and independently, the direct observation of gravitational waves. The results obtained so far augur well, and new observations may be expected before the end of the century.

The phenomena predicted are not specific to general relativity alone, but observation of them would provide new tests. The direct observation of gravitational waves would also found a new astronomy which would add greatly to our knowledge of the Universe, its contents, the phenomena within it and their mechanisms.

9.1 The gyroscope

9.1.1 Evolution equation for classical spin

Consider a system whose centre of mass is G *in Newtonian mechanics*. We introduce a direct orthonormal reference frame R_G centred on G at time t, whose axes point in fixed directions. The spin $\vec{s}$ of the system is then defined by its angular momentum with respect to G, with the speeds of the matter particles being measured with respect to R_G. Such a system constitutes a gyroscope. We assume that the gyroscope is suspended in such a way that the

The figure below shows a schematic picture of VIRGO, the Franco-Italian interferometric gravitational wave detector; construction of this has started (in spring 1996) at Cascina near Pisa.

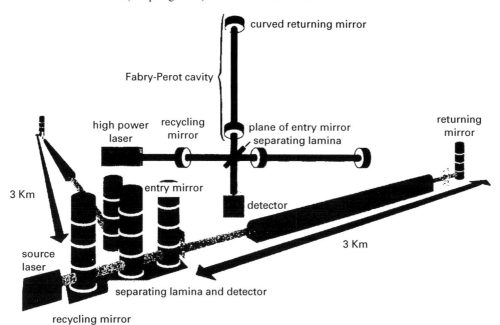

Fig. 9.1

The arms of the interferometer are Fabry–Perot cavities in which the light on average runs back and forth many times before emerging, increasing the sensitivity of the apparatus described in § 9.2.3. Further, under conditions of optimal sensitivity the apparatus works in such a way that the light intensity falling on the detector is very weak. Light thus emerges mainly in the direction of the laser source. The reader should show that the sensitivity decreases with the light intensity in the interferometer. Thus a semi-transparent recycling mirror is used which returns, in phase, to the interferometer light which would otherwise be lost. Vacuum tubes three kilometres long and about one metre in diameter allow the light to travel without disturbance. Towers about 11 metres high contain seismic insulation systems. The seismic insulation has been developed in Pisa, at INFN (Istituto Nazionale di Fisica Nucleare) under the direction of A. Giazotto, while an experimental group directed by A. Brillet in Orsay at CNRS (Centre National de la Recherche Scientifique) is undertaking the optical development. Several other laboratories are involved in France and in Italy.

LIGO is an American project, similar to VIRGO using two interferometers and GEO is an interferometer being developed by German and British Groups.

(VIRGO/LAL)

moment of the forces with respect to G remains zero whatever acceleration it feels. The spin therefore has constant components relative to the frame R_G. We write its components as s^k, with $k = 1, 2$ or 3.

To generalise the notion of spin *in general relativity* we identify the frame R_G with a rest frame (cf. I § 4.1.1) and we introduce the spin four-vector S^α whose components S^k in the rest frame are equal to the components s^k while S^0 vanishes. This definition gives the relation

♦
$$S^\alpha U_\alpha = 0 \qquad (9.1.1)$$

where U^α is the four-velocity of the gyroscope. The relation found here is manifestly covariant, and is therefore satisfied whatever the coordinate system.

We consider an event E on the worldline of the gyroscope, corresponding to the proper time τ measured on the worldline from an arbitrary origin. We introduce the corresponding tangent frame $R_{(\tau)}$.

In $R_{(\tau)}$ the four-velocity and spin have components $\{U^0, U^k\} = \{c, 0, 0, 0\}$ and $\{S^\alpha_{(\tau)}\} = \{S^0, S^k\} = \{0, s^k\}$ respectively. In the same frame $R_{(\tau)}$ the four-velocity and the spin at $\tau + d\tau$ have components $\{U^\alpha_{(\tau+d\tau)}\}$ and $\{S^\alpha_{(\tau+d\tau)}\}$.

At E, the velocity of the gyroscope with respect to this frame is zero; it remains small for events near E. In the slow-motion approximation, i.e. if $d\tau$ is small enough, we assume the Newtonian law of evolution between τ and $\tau + d\tau$, which implies $S^k_{(\tau+d\tau)} = s^k = S^k$. In the tangent frame $R_{(\tau)}$ the acceleration at E has components $\{2a, 0, 0\}$; in this frame the four-velocity at $\tau + d\tau$ is thus $\{U^\alpha_{(\tau+d\tau)}\} = \{U^0_{(\tau+d\tau)}, 2ad\tau, 0, 0\}$ to the second order in $d\tau$. Using the normalisation of four-velocities, $(U^\alpha U_\alpha)_{(\tau+d\tau)} = c^2$, we find the expression $U^0_{(\tau+d\tau)} = c = U^0_{(\tau)}$ up to the second order in $d\tau$. The relation (9.1.1) then allows us to find $S^0_{(\tau+d\tau)}$ to the same order:

$$dS^0 \equiv S^0_{(\tau+\delta\tau)} - S^0_{(\tau)} = \frac{2ad\tau}{c} S^1_{(\tau)}, \qquad dS^k = 0.$$

It is possible to express dS^α in a manifestly covariant form by introducing the four-acceleration Γ^μ whose components in $R_{(\tau)}$ are $\{\Gamma^\mu\} = \{0, 2a, 0, 0\}$:

$$\nabla S^\mu = -\frac{1}{c^2} \Gamma^\rho S_\rho U^\mu d\tau. \qquad (9.1.2)$$

This expression, which can be checked at E in $R_{(\tau)}$, where $\nabla S^\mu = dS^\mu$, expresses the equality between two vectors; it then retains the same form in every frame, in particular the natural frames associated with the coordinate system being used.

Using the relation (9.1.1), $U^\alpha S_\alpha = 0$, equation (9.1.2) can be written

♦
$$\frac{\nabla S^\mu}{d\tau} \equiv \frac{dS^\mu}{d\tau} + \Gamma^\mu_{\ \alpha\beta} S^\alpha U^\beta = \frac{1}{c^2} (\Gamma^\mu U^\rho - \Gamma^\rho U^\mu) S_\rho \qquad (9.1.3)$$

where $\Gamma^\mu_{\ \alpha\beta}$ are the connection coefficients.

Equation (9.1.3) describes the evolution of a classical spin. The relation implies that $S^\alpha U_\alpha$ and $S^\alpha S_\alpha$ remain constant; thus equation (9.1.1), $S^\alpha U_\alpha = 0$, need hold only at one point for it to hold everywhere on the worldline of the gyroscope without the necessity of postulating it everywhere.

$S^\alpha U_\alpha$ is a scalar, and we deduce the relation

$$\frac{d}{d\tau}(S^\alpha U_\alpha) = \frac{\nabla S^\alpha}{d\tau} U_\alpha + S^\alpha \frac{\nabla U_\alpha}{d\tau}, \quad \text{with} \quad \frac{\nabla U_\alpha}{d\tau} \equiv \Gamma_\alpha, \quad \text{and} \quad \Gamma_\alpha U^\alpha = 0.$$

Using (9.1.3) to eliminate $\nabla S^\alpha/d\tau$, we thus verify that $d(S^\alpha U_\alpha)/d\tau = 0$.

Consider the motion of a gyroscope in special relativity, given in a system of Minkowski coordinates in the form $x^k = f^k(t)$, with $ct = x^0$. Let V^k be the velocity of the gyroscope, $V^k = dx^k/dt$, and a^k the acceleration, $a^k = dV^k/dt$. We assume the velocity small enough that we can neglect terms of order $(V^k/c)^3$ or higher. Then the relations $U^\alpha U_\alpha = c^2$ and $U^k = U^0 V^k/c$ give U^0 as a function of V^k. The orthogonality relation $S^\alpha U_\alpha = 0$ gives S^0. To find $\Gamma^k = dU^k/d\tau$ we use the relation $cd(\ )/d\tau = U^0 \cdot d(\ )/dt$ and the above relations. The orthogonality relations $\Gamma^\mu U_\mu = 0$ give Γ^0. We thus get the following expressions, in which we use the vector notation of three-dimensional Euclidean geometry:

$$U^0 = c\left(1 + \frac{\vec{V}^2}{2c^2}\right), \quad S^0 = \vec{S} \cdot \frac{\vec{V}}{c}, \quad \vec{\Gamma} = \dot{a} + \left(\frac{\vec{V}}{c} \cdot \dot{a}\right)\frac{\vec{V}}{c} + \left(\frac{\vec{V}}{c}\right)^2 \dot{a},$$

$$\Gamma^0 = \vec{\Gamma} \cdot \frac{\vec{V}}{c} = \dot{a} \cdot \frac{\vec{V}}{c}$$

with $\vec{V} = \{V^k\}$, $\vec{\Gamma} = \{\Gamma^k\}$, $\dot{a} = \frac{d\vec{V}}{dt}$, $\vec{S} = \{S^k\}$.

The evolution law (9.1.3) for the spin becomes explicitly

$$\frac{d\vec{S}}{dt} = \left(\frac{\dot{a}}{c} \cdot \vec{S}\right)\frac{\vec{V}}{c}, \quad \frac{dS^0}{dt} = \left(1 + \frac{\vec{V}^2}{c^2}\right)\left(\frac{\dot{a}}{c} \cdot \vec{S}\right).$$

One usually introduces the rest frame of the gyroscope which can be found from the original frame through a Lorentz transformation called a **boost**. Neglecting terms in $(V^k/c)^3$, the corresponding rest-frame is defined by the four basis vectors $\vec{e}_{(\alpha)} = e^\mu_{(\alpha)} \vec{\partial}_\mu$:

$$e^\mu_{(0)} = \frac{U^\mu}{c} : \quad e^0_{(0)} = 1 + \frac{\vec{V}^2}{2c^2}, \quad e^k_{(0)} = \frac{V^k}{c}$$

$$e^0_{(k)} = \frac{V^k}{c}, \quad e^j_{(k)} = \delta^j_k - \frac{V^j V_k}{2c^2} = \delta^j_k + \frac{V^j V^k}{2c^2}.$$

We can verify that the new frame is Minkowskian. The reader should show that the *boost* gives the original basis vectors $\vec{\partial}_\mu$ starting from the $\vec{e}_{(\alpha)}$ as given above, with the substitution of V^k for $-V^k$. Note also that the boost reduces to a special Lorentz transformation, to the accuracy considered, for $V^2 = V^3 = 0$.

The orthogonality relation (9.1.1) implies that the spin is a purely spacelike vector in the new frame.

The components s^k of the spin in the frame $\{\vec{e}_{(\alpha)}\}$ have the expressions $s^k = -s_k = -e^\mu_{(k)} S_\mu$. To the approximation considered we find the following relations:

$$\{s^k\} = \vec{s} = \vec{S} - \frac{1}{2}\frac{\vec{V}}{c}\left(\vec{S} \cdot \frac{\vec{V}}{c}\right) \Rightarrow \vec{S} = \vec{s} + \frac{1}{2}\frac{\vec{V}}{c}\left(\vec{s} \cdot \frac{\vec{V}}{c}\right)$$

$$\frac{d\vec{s}}{d\tau} = \left(1 + \frac{\vec{V}^2}{2c^2}\right)\frac{d\vec{s}}{dt} = \frac{1}{2}\left(\left(\frac{\dot{a}}{c}\cdot\vec{S}\right)\frac{\vec{V}}{c} - \left(\vec{S}\cdot\frac{\vec{V}}{c}\right)\frac{\dot{a}}{c}\right) = \vec{\Omega}\wedge\vec{s}$$

with $\vec{\Omega} = \frac{1}{2}\dot{a}\wedge\vec{V}/c^2$.

If the gyroscope has a periodic motion the tetrad $\{\vec{e}_{(\alpha)}\}$ is the same after one period. However, the spin precesses with angular velocity $\vec{\Omega}$ with respect to the tetrad; it therefore does not return to its original value. This is called the **Thomas precession**, and plays an important role in explaining the fine structure of the hydrogen atom. Here the gyroscope is an electron and its quantum spin.

In the presence of gravitation the laws governing the motion of the spin retain the form (9.1.3) with the initial condition (9.1.1) which is always satisfied if we take account of (9.1.3) (cf. II § 6.1). It is, however, impossible to make the connection vanish everywhere by the suitable choice of a global coordinate system.

Given a space-time curve, vectors like S^α which are transported along the curve according to (9.1.3) are said to be **Fermi–Walker transported**.

As an exercise verify that U^α is Fermi–Walker transported and that the scalar product of two vectors is preserved under this transport. Verify also that Fermi–Walker transport reduces to parallel transport if the worldline is a geodesic.

9.1.2 Gravitational field of a rotating body

Consider in Newtonian mechanics a system of orthonormal Galilean coordinates, whose origin O is the centre of a sphere rotating about the axis Oz with angular velocity ω. We assume that the density ρ of the sphere is constant, and we write r_0 for its radius. The kinetic energy density is of order $\rho\omega^2 r_0^2$, much less than ρc^2, and of the same order as the stresses keeping the sphere together. We assume that ω is small enough that we can neglect this term of order two or more. With these assumptions, identifying the Galilean coordinate system considered with a Minkowski system in special relativity, the sphere is described to a first approximation by the energy–momentum tensor $T^{\alpha\beta}$ with

$$T^{00} = \rho c^2 \quad T^{0k} = T^{k0} = \rho c V^k \quad T^{kj} = 0$$

where $\{V^k\} = \{-\omega y, \omega x, 0\}$ represents the Newtonian velocity of the point with coordinates x, y, and z.

We can determine the gravitational field $\phi^{\alpha\beta}$ in the weak-field approximation using the result (8.3.2) of the last chapter:

$$\phi^{00} = -\frac{\chi M c^2}{4\pi r}, \quad \phi^{kj} = \frac{\chi M c^2}{4\pi r}\eta^{kj}, \quad \phi^{0k} = \phi^{k0} = -\frac{\chi}{2\pi}\iiint\frac{\rho c V^k}{R}d^3r'.$$

The total mass of the sphere is M, and $r = \sqrt{x^2 + y^2 + z^2}$ and $R = \sqrt{(\vec{r} - \vec{r}')^2}$. We note that $\phi^{\alpha\beta}$ is independent of t, and its expression is given for the point $\vec{r} = \{x, y, z\}$ although in the integral V^k is expressed at the point $\vec{r}\,'$. The integration is over the interior of the sphere, where ρ is non-zero.

We now assume that r is large enough that we need retain only the first non-zero term of the expansion of ϕ^{k0} in powers of $1/r$. This term is of order r^{-2}:

$$\phi^{0k} = \phi^{k0} = -\frac{\chi\rho c}{2\pi r^3} \iiint V^k \vec{r} \cdot \vec{r}\,' d^3 r'. \tag{9.1.4}$$

Using the expression for V^k, $\{V^k\} = \{-\omega y, \omega x, 0\}$, we integrate (9.1.4):

$$\phi^{01} = \frac{\chi c}{4\pi r^3} Jy, \quad \phi^{02} = -\frac{\chi c}{4\pi r^3} Jx, \quad \phi^{03} = 0, \quad \text{with } J = \frac{8\pi}{15}\rho r_0^5 \omega.$$

J is the component along Oz of the angular momentum $\vec{J}$ of the rotating sphere: $\vec{J} = J\vec{k}$ where $\vec{k}$ is the unit vector along Oz. The metric may then be expressed in the following form:

$$\blacklozenge \quad \begin{cases} g_{\alpha\beta} = \eta_{\alpha\beta} + h_{\alpha\beta}, \text{ with } h_{\alpha\beta} = \phi_{\alpha\beta} \\[2mm] U \equiv \dfrac{-GM}{rc^2}, \quad \vec{h} \equiv \{h_{0k}\} \\[2mm] h_{00} = 2U, \quad h_{jk} = -2\gamma U \eta_{jk}, \quad \vec{h} = \dfrac{2G}{c^3 r^3}\vec{J} \wedge \vec{r}, \quad \gamma = 1. \end{cases} \tag{9.1.5}$$

The coefficient γ is introduced here only so that we can recognise expressions depending on U in the final results.

The reader should compare the approximations made here with those of the preceding chapter.

The hypothesis of spherical symmetry leads to the above expressions for ϕ^{00} and ϕ^{kj} to first order in χ and avoids the need to expand these terms to order r^{-2}. An expansion to that order would be necessary in the absence of spherical symmetry. The presence of terms in r^{-2} in h_{00} leads to Newtonian effects of the same type as the relativistic term of order χ^2 which appears in the expansion of the Schwarzschild metric in isotropic coordinates (II § 7.2.2). Thus the precession of the perihelion of Mercury could be partly explained in a Newtonian framework by an oblateness of the Sun. This possibility gave rise to several controversies in the past; some authors considered that the perihelion precession was essentially explained by solar oblateness, thus challenging the validity of general relativity. The situation has evolved. In the post-Newtonian approximation one tries today to use general relativity wherever possible, rather than to test it, as the theory is well established (cf. § 3.4.4).

The reader should study the precession of the orbits of satellites about a central body in the equatorial plane $z = 0$, and discuss the effect of h_{0k}.

9.1.3 Observation of distant stars

The relations above determine the metric in a coordinate system that we should like to interpret physically. Giving a physical interpretation to a coordinate system is not absolutely necessary if we control every element of the experimental apparatus we use; however, if we wish to use the distant stars in our observations, we have to make some hypotheses about their apparent behaviour as seen by an observer who is at rest in the coordinate system.

In Newtonian theory we assume that Galilean reference frames preserve a fixed direction with respect to the distant stars. This assumption is compatible with observations of their apparent relative motion which is for the most part negligible. A justification of this hypothesis is to be sought *a posteriori* in the harmony between observations, experimental results and theoretical predictions. The situation is analogous here.

Neglecting the rotation of the central body, we know that we get the Schwarzschild metric in isotropic coordinates to first order in χ (II § 7.2.2). When the central mass is the Sun we have identified these coordinates with Galilean coordinates in Newtonian theory. With this intepretation, in the slow-motion limit we recover Newtonian theory, all contained in the expression for h_{00}, by setting $\gamma = 0$ (cf. II § 2.4.2). To order χ this interpretation provides an explanation for the deflection of light rays by the Sun, and to order χ^2 for the precession of planetary perihelia (cf. II § 3.4 and 7.2.2). The agreement between the theoretically predicted values and those observed justifies the interpretation.

For the study of relativistic perturbations in Newtonian theory, the coordinate system may be regarded as a Galilean system, the relativistic laws having only the role of determining corrections to Newtonian physics. Thus to express the apparent angle between two distant stars in terms of their angular coordinate difference requires us to take account of the deflection of light rays by the Sun if this is non-negligible. The calculation procedure which gives the coordinates from the observations is therefore not exactly the Newtonian procedure. This point of view, giving an *a priori* meaning to the coordinate system, robs the theory of one of its distinctive characteristics. It can only be used with great caution, and then only in the quasi-Newtonian case. Nevertheless this is a simple way of using physical intuition based on understanding of Newtonian theory in the "post-Newtonian" approximation used here.

We adopt the previous interpretation of the coordinate system; we assume in particular that the distant stars retain fixed apparent directions for any "static" observer. We still have to make precise the idea of a "static observer".

Consider a material point static at the point N with arbitrary but constant spatial coordinates. At N we introduce at each time four vectors $\vec{h}_{(\alpha)} = h^{\mu}{}_{(\alpha)} \vec{\partial}_{\mu}$ where $\{\vec{\partial}_{\mu}\}$ is the natural basis associated with the coordinate system. These four vectors are numbered by the index (α). They are chosen so that:

(1) the four vectors satisfy the relations $g_{\mu\nu} h^{\mu}{}_{(\alpha)} h^{\nu}{}_{(\beta)} = \eta_{\alpha\beta}$ – vectors satisfying these orthonormality relations form a **tetrad**;

(2) the coefficients $h^{\mu}{}_{(\alpha)}$ are independent of time; we then consider that the tetrad thus defined at N at each time does not rotate with respect to the natural frame; as the metric is time-independent the existence of such tetrads is assured;

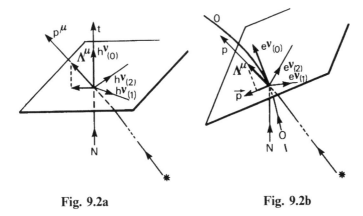

Fig. 9.2a   Fig. 9.2b

(3) $ch^\mu{}_{(0)}$ is the four-velocity of a material point at rest at N, which implies $h^\mu{}_{(0)} = \delta^\mu{}_0/\sqrt{g_{00}}$.

There exist several tetrads satisfying these conditions. As an exercise the reader should find these to first order in $h_{\alpha\beta}$.

We define a "static observer" by giving point N and one of the above tetrads which allows us to specify four-vectors defined at N by their components relative to the basis $\{h^\mu{}_{(\alpha)}\}$. The vector space spanned by the three vectors $\{h^\mu{}_{(k)}\}$ is orthogonal to the observer's four-velocity, so that for this observer it constitutes the set of purely spatial vectors.

Let P^μ be the components in the natural basis of the four-momentum of a photon coming from a distant star, reaching the static observer at N at the instant t (fig. 9.2a). The four-momentum of the photon is a null vector with components $\ell\{1, \ell^k\}$ in the frame formed by the tetrad, with $-\eta_{jk}\ell^j\ell^k = \sum_k \ell^k\ell^k = 1$. For the static observer, the unit spacelike vector $-\vec{\ell} = -\{0, \ell^k\}$ defines the apparent direction of the star, while ℓc is the photon energy for this observer.

We will say that the distant star preserves a fixed apparent direction for the static observer at N if ℓ^k is independent of the time of observation. Our hypothesis here is that for every static observer, the distant stars retain fixed directions.

It is easy to find P^μ in the form $P^\mu = \ell(h^\mu{}_{(0)} + h^\mu{}_{(k)}\ell^k)$. The tetrad coefficients $h^\mu{}_{(\alpha)}$ are constant. The worldline of photons emitted by a distant star is thus tangent at N to the vector Λ^μ, $\Lambda^\mu = (h^\mu{}_{(0)} + h^\mu{}_{(k)}\ell^k)$, with constant components in the natural frame and independent of the photon energy. The apparent direction of the star is determined by the projection of Λ^μ on the vector subspace spanned by $h^\mu{}_{(k)}$.

Analogously, we can consider a second observer O, at N, at time t. This observer is not necessarily static: he crosses N with four-velocity U^μ. As before, we associate with this oberver a tetrad $e^\mu{}_{(\alpha)}$ such that $U^\mu = ce^\mu{}_{(0)}$ (fig.

9.2b). The vectors $e^\mu{}_{(k)}$, orthogonal to U^μ, constitute an orthonormal basis for the "purely spacelike" vectors of the second observer at time t. Photons coming from the star have four-momentum with components $P^\mu = \ell \Lambda^\mu$ in the natural basis. The tetrad components of this vector are $\ell \Lambda_\mu e^\mu{}_{(\alpha)} \eta^{\alpha\beta} = p\{1, p^k\}$, with $\sum_k p^k p^k = 1$. The apparent direction of the star for the second observer O is given by the vector $-\vec{p} = -\{0, p^k\}$; $\vec{p}$ is the direction of the projection of the vector Λ^μ onto the subspace spanned by the space vectors $e^\mu{}_{(k)}$.

> Consider two stars characterised by the vectors $\Lambda^\mu{}_1$ and $\Lambda^\mu{}_2$. For the moving observer the angle θ_e between the two stars is given by $\cos \theta_e = \vec{p_1} \cdot \vec{p_2}$, while for the static observer the apparent angle between the two stars is θ_h, with $\cos \theta_h = \vec{\ell_1} \cdot \vec{\ell_2}$. These two angles are in general different, which is the phenomenon of **aberration**. As an exercise the reader should find θ_h and θ_e in the tetrad formalism when $\vec{\ell_1} = \{1, 0, 0\}$, $\vec{\ell_2} = \{0, 1, 0\}$ and $U^\mu = U^{(\alpha)} h^\mu{}_{(\alpha)}$. Choose $U^{(\alpha)}$ in the form $\{U^{(\alpha)}\} = f\{c, V, 0, 0\}$. Express f as a function of V, and make precise what is meant by V.
>
> Study the equations of null geodesics, photon worldlines, and show that the spatial trajectories of photons reaching N from a fixed apparent direction are identical.

Assume that the second observer returns regularly to N, with the same four-velocity U^μ. It is then possible always to associate the "same" tetrad with the observer whenever he is at N; i.e. always to choose the same values for $e^\mu{}_{(\alpha)}$. In this case the components p^k take the same values: the stars have the same apparent direction with respect to $e^\mu{}_{(k)}$.

Assume that the observer moves periodically. At a given arbitrary point of the trajectory, U^μ always has the same value. In this case the observer can choose a tetrad characterised by coefficients $e^\mu{}_{(\alpha)}$ which only depend on his position in space. This does not mean that $e^\mu{}_{(\alpha)}$ is time-independent nor that the axes pointing along $e^\mu{}_{(k)}$ always necessarily point to the same stars; it means only that the vectors $e^\mu{}_{(k)}$ point towards the same stars every time the observer returns to the same point.

Assume that a gyroscope carried by the observer has a rotation with respect to the vectors $e^\mu{}_{(k)}$ such that the spin components do not return to their previous values when the gyroscope returns to the same point N. In this case, which we shall consider again below, the gyroscope precesses with respect to the distant stars. This motion is directly observable by comparing the direction of the gyroscope's spin with that of telescopes which point towards selected stars. The expected effect is very small but is cumulative: this is called a "secular" effect.

9.1.4 Motion of a satellite gyroscope

Consider a gyroscope carried in a satellite around a rotating central mass. As the gyroscope is freely falling, its acceleration Γ^μ vanishes. Equation (9.1.3) becomes

$$\frac{dS^\mu}{d\tau} + \Gamma^\mu{}_{\alpha\beta} S^\alpha U^\beta = 0. \tag{9.1.6}$$

The spin of the gyroscope is thus parallel-transported.

In agreement with the interpretation of the coordinate system, we use vector *notations* from Euclidean geometry, obtained by regarding the hypersurfaces $t = $ constant as the absolute space of Newtonian theory. The motion of the gyroscope is characterised by the three functions $x^k = f^k(t)$. We set

$$\vec{r} = \{x^k\}, \ \vec{V} = \{V^k\} = \left\{\frac{df^k}{dt}\right\}, \ \vec{S} = \{S^k\}, \ \vec{h} = \{h_{0k}\}.$$

In the metric (9.1.5), following the discussion of chapter 2 (II section 2.3), the orbital motion of the gyroscope can be derived from the Lagrangian L:

$$L = \sqrt{1 + 2U + 2\vec{h}\cdot\vec{V}/c - (1 - 2\gamma U)\vec{V}^2/c^2}. \tag{9.1.7}$$

As the motion is Newtonian to first approximation, the kinetic energy is of the same order of magnitude as the gravitational potential energy. We may thus neglect terms of order $U(V/c)^2$ in the expansion of L, as they are comparable with the terms of order U^2 we have already assumed negligible in linearised theory. We further assume that ω is small enough for us to neglect terms of order $h_{0k}(V/c)$. Under these conditions, expanding the Lagrangian L we see that the orbital motion of the gyroscope is that predicted in Newtonian theory, even if we include terms of order $(V/c)^3$ and $U(V/c)$. Terms of this order of magnitude do however affect the spin of a satellite gyroscope, even though they do not influence its orbital motion.

In making approximations it is useful to substitute εV^k for V^k and $\varepsilon^2 U$ for U while h_{0k} is replaced by ηh_{0k} in the expression (9.1.5) for the metric and in the evolution law (9.1.6) for the spin. The parameters ε and η are regarded as infinitesimal; the expansions are limited to terms of order less than ε^4, η^2 and $\varepsilon\eta$. We set $\varepsilon = \eta = 1$ at the end of the calculation.

> We have assumed that the spin does not influence the motion of the gyroscope, which remains that of a test particle. This approximation can be justified here. In contrast, the acceleration effects leading to the Thomas precession must be eliminated; thus we have to protect the satellite from perturbations such as those of the solar wind. For this reason the gyroscope is protected by an outer envelope whose motion is slaved to the free-fall motion of the gyroscope which takes place inside it.

To describe the motion of the spin one usually introduces a comoving observer at each point of the gyroscope's worldline, with associated tetrad

$e^\mu{}_{(\alpha)}$:

$$
\begin{cases}
e^\mu{}_{(0)} = \dfrac{U^\mu}{c} : \quad e^0{}_{(0)} = 1 - U + \dfrac{\vec{V}^2}{2c^2}, \quad e^k{}_{(0)} = \dfrac{V^k}{c} + O(\varepsilon^3) \\[2mm]
e^0{}_{(k)} = \dfrac{V^k}{c} + O(\varepsilon^3, \eta) \\[2mm]
e^j{}_{(k)} = (1 + \gamma U)\delta^j{}_k - \dfrac{V^j V_k}{2c^2} = (1 + \gamma U)\delta^j{}_k + \dfrac{V^j V^k}{2c^2}.
\end{cases}
\tag{9.1.8}
$$

In these expressions the various quantities are calculated at the space–time point considered on the gyroscope's worldline. We assume that equalities are satisfied to terms of order ε^4, η^2 and $\varepsilon\eta$; however, we denote by $O(\varepsilon^3, \eta)$ the presence of terms of order ε^3 and η which are not *a priori* negligible but whose explicit value is not necessary for the calculation.

> The reader should show that the frame introduced constitutes a suitable tetrad and find expressions for the terms $O(\varepsilon^3, \eta)$ and $O(\varepsilon^3)$ which are not given explicitly in (9.1.8).

Let $s_{(\mu)}$ and $s^{(\nu)}$ be the covariant and contravariant components of the spin in the tetrad (9.1.8). The formulae for change of basis give

$$
s_{(\mu)} = e^\alpha{}_{(\mu)} S_\alpha = e^\alpha{}_{(\mu)} g_{\alpha\beta} S^\beta \quad \text{and} \quad S^\beta = e^\beta{}_{(\mu)} s^{(\mu)}
$$

where S_α and S^β are the covariant and contravariant spin components in the initial frame, the natural frame of the coordinate system at the space-time point considered on the gyroscope's worldline. The indices α and β are lowered and raised using the metric $g_{\alpha\beta}$ or its inverse $g^{\alpha\beta}$. As the frame $\{e^\mu{}_{(\alpha)}\}$ is Minkowskian the indices in parentheses, (μ), are lowered or raised using $\eta_{\alpha\beta}$ or $\eta^{\alpha\beta}$.

We may regard $s_{(\mu)} = e^\beta{}_{(\mu)} S_\beta$ as the four-dimensional scalar product of the vectors $e^\beta{}_{(\mu)}$ and S^α. It is therefore a scalar whose ordinary derivative along the gyroscope's worldline is equal to the covariant derivative. The spin evolution law (9.1.6) shows that the covariant derivative of S^α and thus of S_β vanishes. The evolution law for $s^{(\nu)}$ is thus:

$$
\frac{ds_{(\mu)}}{d\tau} = \frac{\nabla s_{(\mu)}}{d\tau} = S_\beta \frac{\nabla e^\beta{}_{(\mu)}}{d\tau} = e^\alpha{}_{(\rho)} s^{(\rho)} g_{\alpha\beta} \frac{\nabla e^\beta{}_{(\mu)}}{d\tau}
\tag{9.1.9}
$$

with $\dfrac{ds^{(\nu)}}{d\tau} = \dfrac{d}{d\tau}\left(\eta^{\nu\mu} s_{(\mu)}\right) = \eta^{\nu\mu} \dfrac{ds_{(\mu)}}{d\tau}$, $\eta^{\nu\mu} = \text{diag}\,(1, -1, -1, -1) = \text{constant}$.

In these expressions the indices in parentheses, (μ), (ρ) and (ν), cannot be regarded as tensor indices but as numbers labelling the four vectors $e^\alpha{}_{(\mu)}$ and the four scalars $e^\alpha{}_{(\mu)} g_{\alpha\beta} S^\beta$; the latter also represent the covariant components of the spin in the basis $\{e^\alpha{}_{(\mu)}\}$.

Evaluating equations (9.1.9) we get the evolution equation

$$
\blacklozenge \qquad \frac{d\vec{s}}{d\tau} = \vec{\Omega} \wedge \vec{s}, \quad \vec{\Omega} = \left((1 + \gamma)\overrightarrow{\nabla U} + \frac{1}{2c^2}\frac{d\vec{V}}{dt}\right) \wedge \vec{V} + \frac{c}{2}\vec{\nabla} \wedge \vec{h}
\tag{9.1.10}
$$

with the notation $\vec{s} = \{s^{(k)}\}$, $\vec{h} = \{h_{0k}\}$, $\vec{\nabla} = \{\partial_k\}$.

The equations of free-fall orbital motion for the gyroscope are the Newtonian equations in the approximation considered; we thus get

$$\blacklozenge \quad \frac{d\vec{V}}{dt} = -c^2 \vec{\nabla} U \Rightarrow \vec{\Omega} = \left(\gamma + \frac{1}{2}\right)\vec{\nabla} U \wedge \vec{V} + \frac{c}{2}\vec{\nabla} \wedge \vec{h}, \quad \gamma = 1. \quad (9.1.11)$$

Equations (9.1.10) show that the spin precesses with angular velocity $\vec{\Omega}$ with respect to the rest frame $\{e^{\alpha}{}_{(\mu)}\}$ of the gyroscope.

In the expression (9.1.11) for $\vec{\Omega}$, the first term gives the **geodetic precession** (or de Sitter precession). This spin–orbit coupling term appears because the gyroscope moves in a gravitational field. We note here the influence of the term in γ, which does not appear in the equations of orbital motion to the approximation considered. In Newtonian language $d\vec{V}/dt$ behaves like the gravitational acceleration of the gyroscope in the initial coordinate system; the expression (9.1.10) shows the resulting Thomas precession. It is, however, interesting to note the difference in the treatment of the acceleration Γ^{μ}, in the initial equations (9.1.3), and the treatment of the gravitational acceleration implied by the connection.

The second term in the expression (9.1.11) for $\vec{\Omega}$ gives the **Lense–Thirring effect**. This effect only depends on the position of the gyroscope and not on its speed; it results from the rotation of the central body.

In a frame moving with angular velocity $\vec{\Omega}$ with respect to the tetrads we have introduced, accelerometers will measure zero acceleration as the frame is freely falling. Moreover, suitably suspended gyroscopes do not precess. This is a "local inertial frame". The Lense–Thirring effect is sometimes called "dragging of inertial frames"; dragging of the local inertial frame by the rotation of the central body.

The gyroscope returns periodically to the same points of its trajectory, points with the same spatial coordinates in the initial coordinate system. The tetrad coefficients $e^{\mu}{}_{(\alpha)}$ depend only on the point N considered on the trajectory as the motion is periodic. We recover here the conditions of the discussion in the previous section.

As it described its orbit the gyroscope underwent a precession with respect to the tetrads. After one revolution the proper time of the gyroscope has increased from τ_1 to τ_2; the spin has changed from $\delta\vec{s}$: it has rotated through an angle $\delta\vec{\theta}$ with respect to the tetrad, i.e. with respect to the distant stars. To lowest order we get

$$\delta\vec{s} = \delta\vec{\theta} \wedge \vec{s}, \quad \delta\vec{\theta} = \int_{\tau_1}^{\tau_2} \vec{\Omega}\, d\tau \equiv \langle\vec{\Omega}\rangle(\tau_2 - \tau_1)$$

where $\vec{\Omega}$ is regarded as a function of proper time τ. The average speed of the precession of the gyroscope with respect to the distant stars is thus $\langle\vec{\Omega}\rangle$.

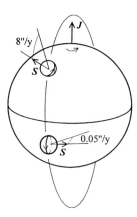

Fig. 9.3

As an exercise, investigate a system in which two satellite gyroscopes on different orbits meet periodically. We assume that all parts of this experiment can be controlled. Show that the direct comparison of the spin directions of the gyroscopes when the satellites are close shows the effects of precession, without the necessity of making any hypothesis at all about the apparent motion of the distant stars.

The Earth is not a perfect sphere, nor an isolated body; its centre is not static in a Galilean reference frame; however, the preceding discussion shows that we can calculate the speed of geodetic precession $\vec{\Omega}_G$ and the speed of the Lense–Thirring precession $\vec{\Omega}_{L-T}$ of terrestrial origin. These angular velocities add to the other contributions arising in various ways.

The average precession rate of the spin of a satellite gyroscope in Earth orbit depends on the orbit; even for a low orbit Ω_G and Ω_{L-T} are very small: less than about ten seconds of arc per year for Ω_G and about one hundred times smaller for Ω_{L-T}. The results found in the laboratory, at Stanford University, currently justify placing such an experiment in orbit which would demonstrate a new effect and provide a further test allowing us to determine the γ parameter of Eddington and Robertson (cf. II § 3.4.1) to an accuracy which could exceed 10^{-4}.

The reader may study the geodetic and Lense–Thirring precessions for low circular orbits whose radii are similar to that of the Earth, i.e. 6400 km. In particular, consider a polar and equatorial orbit for various initial orientations of the spin, in the plane of the orbit and perpendicular to it. Recall that the mass of the Earth is of order 6×10^{24} kg and that it rotates once per 24 hours.

9.2 Gravitational waves

We have seen in the previous chapter that general relativity predicts the existence of gravitational waves. These are emitted by accelerated masses, in a similar way to the emission of electromagnetic waves by accelerated charges in Maxwell's theory. Under certain conditions, such as weak gravitational fields and slow motion, which we assume to hold here, gravitational waves can be described in Newtonian terms.

We shall use Newtonian language as much as possible, with the following notation: $x = \{ct, \vec{r}\} = \{x^0, r^k\}$, $k = 1, 2$ or 3. However, we shall lower or raise Latin indices j, k, a, b, etc., using η_{jk} or its inverse η^{jk}: $\eta_{jk} = \mathrm{diag}(-1,-1,-1)$; we also use the summation convention for repeated indices. As usual, G denotes the Newtonian gravitational constant and c the speed of light in vacuo.

Gravitational waves are then characterised in a Galilean coordinate system by a field $h^{(W)}{}_{jk}(x)$. Far from the source, the source's reduced quadrupole moment $\maltese^{jk}$ alone determines the wave:

$$\maltese^{jk} \equiv \iiint \left(r^j r^k + \frac{1}{3}\vec{r}^{\,2}\eta^{jk} \right) \rho(\vec{r}, t)d^3r \qquad (9.2.1)$$

where ρ is the mass density at the point with space coordinates $\vec{r} = \{r^k\}$. The volume element is written d^3r and the integration is over all space, i.e. the whole region where ρ does not vanish.

Under these conditions, $h^{(W)}{}_{jk}$ takes the form

$$h^{(W)}{}_{jk}(\vec{r}, t) = h^{(W)}{}_{kj} = -\frac{2G}{c^4}\frac{1}{r}\, P_{jkab}\, \maltese^{ab}\left(t - \frac{r}{c}\right) \qquad (9.2.2)$$

with $r = \|\vec{r}\|$, the distance of the source from the observer, assumed to be very large compared with the size of the source, and $(\,\ddot{}\,) = d^2(\ )/dt^2$. We use here the standard notation $\maltese^{ab}$ rather than the more correct notation $\maltese^{ab}$ of the last chapter.

> As an exercise compare the extent of a stellar system of the size of the solar system with the distance between us and the nearest star.

The quantity P_{jkab}, introduced in (9.2.2), is defined as follows:

$$P_{jkab} \equiv \Pi_{ja}\Pi_{kb} - \frac{1}{2}\Pi_{jk}\Pi_{ab} \qquad \Pi_{jk} = \eta_{jk} + n_j n_k \qquad (9.2.3)$$

with $j, k, a, b, \ldots = 1, 2$ or 3 and $n_k = -n^k = -r^k/r$.

We note that $\maltese^{jk}$, η_{jk}, η^{jk}, P_{jkab}, n^j, n_j and $h^{(W)}{}_{jk}$ behave like three-dimensional tensors under a spatial coordinate change.

Gravitational waves (9.2.2) obey d'Alembert's linear wave equation

$$\Box h^{(W)}{}_{jk} = 0 \qquad (9.2.4)$$

far from the source, as well as the transversality relation $n^k h^{(W)}{}_{jk} = 0$.

Far from the source, gravitational waves propagate with speed c, in agreement with (9.2.4) or its solution (9.2.2). The waves carry energy, and the loss of this energy perturbs the motion of a binary system. Observations of this perturbation constitute a positive test of general relativity, with precision close to 3×10^{-3}. We can regard the existence of gravitational waves as indirectly proven by these observations. However, no direct observation confirming their existence has yet been performed as of 1996.

To make direct observations, one needs very sensitive detectors as the expected effects are very small. The first mechanical detectors were built in the 1960s by Joseph Weber at the University of Maryland. In this type of detector gravitational waves cause strains and deformations in a bar, and one attempts to measure these.

Here we shall present a more promising detection principle, based on interferometric methods first studied in the 1970s.

9.2.1 Radiated energy

In special relativity, in chapter 6 of part I, we introduced the energy–momentum tensor $T_{(\text{tot})}{}^{\alpha\beta}$, as the sum of two terms, the first corresponding to the matter distribution, and the second to the electromagnetic field. This tensor obeys $T_{(\text{tot})}{}^{\alpha\beta}{}_{,\beta} = 0$. These relations allow us to define conserved quantities, such as the total energy and momentum of an isolated system for example; they allow us to study energy and momentum transfer between matter and an electromagnetic field, and to keep account of the energy budget when the system radiates.

To have similar possibilities for the gravitational field, we need to define a quantity $\tau^{\alpha\beta}$ which satisfies conservation equations written with ordinary rather than covariant derivatives:

$$\tau^{\alpha\beta}{}_{,\beta} = 0. \tag{9.2.5}$$

In quasi-Minkowski coordinates, when the gravitational field is negligible, the quantities $\tau^{\alpha\beta}$ should be identical with the components of a special-relativistic energy-momentum tensor, so as to maintain the physical interpretation of the notions of energy and momentum.

One can construct several symmetrical quantities $\tau^{\alpha\beta}$ of the form $g(T^{\alpha\beta} + t^{\alpha\beta})$ with the required properties. Here g is the modulus of the determinant of the metric, $g = |\text{Det}[g_{\alpha\beta}]|$, while $T^{\alpha\beta}$ is the energy–momentum tensor of the sources of the gravitational field and $t^{\alpha\beta}$ is a symmetric quantity $t^{\alpha\beta} = t^{\beta\alpha}$, which contains the specific contribution of the gravitational field. In general, $t^{\alpha\beta}$ is not a tensor, and it is called the **energy–momentum pseudo-tensor**.

The quantities τ^{00} and τ^{0k}/c can be interpreted as densities of energy and momentum.

The energy E_V in the volume V containing the matter is defined by

$$E_V = \iiint_V \tau^{00} d^3 r. \tag{9.2.6}$$

In the expression for E_V all forms of energy are taken into account.

The energy E_V is a function of time, t. The conservation relation $\tau^{0\beta}{}_{,\beta} = 0$ implies that the decrease of energy per unit time, $-dE_V/dt$, is the flux through Σ, $L_{G/\Sigma}$, of the vector pseudo-density $\vec{\tau} = c\{\tau^{0k}\}$ which we interpret as the gravitational power radiated across the surface Σ bounding the volume V considered:

$$L_{G/\Sigma} = c \iint_{\Sigma} \sum_k \tau^{0k} n^k d\Sigma = -\frac{dE_V}{dt} \qquad (9.2.7)$$

where $\{n^k\} = \vec{n}$ is the vector normal to Σ, pointing to the exterior of V.

We integrate the relation $\tau^{\alpha\beta}{}_{,\beta} = 0$ over V and use Gauss's theorem; we thus obtain the following relations leading to (9.2.7):

$$\iiint_V \tau^{00}{}_{,0} d^3r + \iiint_V \tau^{0k}{}_{,k} d^3r = 0$$

$$\Rightarrow -\frac{d}{dt} \iiint_V \tau^{00} d^3r = c \iint_{\Sigma} \sum_k \tau^{0k} n^k d\Sigma \equiv L_{G/\Sigma}.$$

The "pseudo-vector" $c\tau^{0k}$ is thus the analogue for gravitation of the Poynting vector in electromagnetism (cf. I § 6.3.4).

The usual picture in which the energy lost by the system is equal to the radiated energy is thus preserved with these definitions.

To construct an energy–momentum pseudo-tensor, we *define* $h_{\alpha\beta}$ exactly, through the relation $g_{\alpha\beta} = \eta_{\alpha b} + h_{\alpha\beta}$. The Einstein tensor appears as the sum of two terms, $S_{(1)}{}^{\alpha\beta}$ and $S_{(2)}{}^{\alpha\beta}$, with the first containing all the terms linear in $h_{\alpha\beta}$ and its derivatives. We multiply both sides of the Einstein equations (6.4.5) by g, $g = |\mathrm{Det}[g_{\alpha\beta}]| \equiv 1 + \delta g$. We get the exact equation

$$S_{(1)}{}^{\alpha\beta} = \chi g T^{\alpha\beta} - g S_{(2)}{}^{\alpha\beta} - \delta g S_{(1)}{}^{\alpha\beta}.$$

We verify easily the identity $S_{(1)}{}^{\alpha\beta}{}_{,\beta} \equiv 0$; from this we can deduce that

$$\left(g T^{\alpha\beta} - \frac{1}{\chi} \left(g S^{\alpha\beta}{}_{(2)} + \delta g S^{\alpha\beta}{}_{(1)} \right) \right)_{,\beta} = 0.$$

We define

$$\tau^{\alpha\beta} \equiv g T^{\alpha\beta} - \frac{1}{\chi} \left(g S^{\alpha\beta}{}_{(2)} + \delta g S^{\alpha\beta}{}_{(1)} \right).$$

$\tau^{\alpha\beta}$ is a symmetrical quantity with the required properties to be regarded as an energy–momentum pseudo-density: $\tau^{\alpha\beta}{}_{,\beta}$ vanishes when the Einstein equations are satisfied, and has the required behaviour in the limit $h_{\alpha\beta} = 0$. It is clear that we can always add to the above expression for $\tau^{\alpha\beta}$ any quantity $X^{\alpha\beta}$ whose divergence $X^{\alpha\beta}{}_{,\beta}$ is identically zero.

We use the harmonic gauge introduced in the previous chapter as well as the notation of the present chapter. We define here $t^{\alpha\beta}$ to first order by the relation:

$$gt^{\alpha\beta} \equiv \frac{1}{4\chi}\left\{ 2\psi^{\alpha}{}_{\sigma,\tau}\psi^{\beta\sigma,\tau} - 2\psi^{\alpha}{}_{\sigma,\tau}\psi^{\sigma\tau,\beta} - 2\psi^{\beta}{}_{\sigma,\tau}\psi^{\sigma\tau,\alpha} + \psi_{\sigma\mu,}{}^{\alpha}\psi^{\sigma\mu,\beta} \right.$$

$$\left. -\frac{1}{2}\psi_{,}{}^{\alpha}\psi_{,}{}^{\beta} - \frac{1}{2}\eta^{\alpha\beta}\left(\psi_{\sigma\mu,\tau}\psi^{\sigma\mu,\tau} - \frac{1}{2}\psi_{,}{}^{\mu}\psi_{,\mu} - 2\psi^{\mu}{}_{\sigma,\tau}\psi^{\sigma\tau}{}_{,\mu}\right)\right\}$$

$$\psi^{\alpha\beta} \equiv \phi^{\alpha\beta} - \frac{1}{2}\eta^{\alpha\beta}\phi^{\mu}{}_{\mu}, \quad \psi \equiv \psi^{\alpha}{}_{\alpha} \Rightarrow \phi^{\alpha\beta} \equiv \psi^{\alpha\beta} - \frac{1}{2}\eta^{\alpha\beta}\psi^{\mu}{}_{\mu}.$$

Further, from (8.2.5), the relation $\psi^{\alpha\beta}{}_{,\beta} = 0$ is satisfied.

To show that $\tau^{\alpha\beta}{}_{,\beta} = (g T^{\alpha\beta} + g t^{\alpha\beta})_{,\beta}$ vanishes we first establish the following identity, which holds when $\psi^{\alpha\beta}{}_{,\beta} = 0$:

$$(g t^{\alpha\beta})_{,\beta} \equiv \frac{1}{4\chi}\left(\phi_{,\beta}\Box\psi^{\alpha\beta} - 2\phi^{\alpha}{}_{\sigma,\beta}\Box\psi^{\sigma\beta} + \phi_{\sigma\beta,}{}^{\alpha}\Box\psi^{\sigma\beta}\right).$$

We can verify this identity to first order by expressing $\phi^{\alpha\beta}$ as a function of $\psi^{\alpha\beta}$. The Einstein equations allow us to express $T^{\alpha\beta}$ as a function of $S^{\alpha\beta}$ in the equations of motion (6.4.4), $\nabla_{\beta}T^{\alpha\beta} = 0$. Using the identity $g_{,\beta} \equiv 2g\Gamma^{\alpha}{}_{\alpha\beta}$, we find

$$\nabla_{\beta}T^{\alpha\beta} \equiv T^{\alpha\beta}{}_{,\beta} + \Gamma^{\alpha}{}_{\sigma\beta}T^{\sigma\beta} + \Gamma^{\beta}{}_{\beta\sigma}T^{\alpha\sigma} = 0$$

$$g\nabla_{\beta}T^{\alpha\beta} = (g T^{\alpha\beta})_{,\beta} + \frac{1}{\chi}\left(g\Gamma^{\alpha}{}_{\sigma\beta}S^{\sigma\beta} - g\Gamma^{\beta}{}_{\sigma\beta}S^{\alpha\sigma}\right) = 0.$$

We expand this relation to first order; χ and $\Gamma^{\alpha}{}_{\sigma\beta}$ are first-order quantities, so it is convenient to replace $S^{\alpha\beta}$ by $S_{(1)}{}^{\alpha\beta}$. We get

$$(g T^{\alpha\beta})_{,\beta} + \frac{1}{4\chi}\left(\phi_{,\beta}\Box\psi^{\alpha\beta} - 2\phi^{\alpha}{}_{\sigma,\beta}\Box\psi^{\sigma\beta} + \phi_{\sigma\beta,}{}^{\alpha}\Box\psi^{\sigma\beta}\right) = 0.$$

This shows that $\tau^{\alpha\beta}{}_{,\beta}$ vanishes if we take account of the identity established at the outset.

When matter is localised near the origin, the "Landau–Lifschitz pseudo-tensor" gives the following expressions for the energy density τ^{00}, and the momentum density τ^{0k}/c, of gravitational radiation (9.2.2) far from the source:

$$\begin{cases} \tau^{00}(\vec{r}, t) = \dfrac{G}{4\pi c^6}\dfrac{1}{r^2}\left(\dfrac{1}{4}\left(n^j n^k \dddot{\mathcal{Y}}_{jk}\right)^2 + \dfrac{1}{2}\dddot{\mathcal{Y}}^{jk}\dddot{\mathcal{Y}}_{jk} + n^j n^k \dddot{\mathcal{Y}}_{jm}\dddot{\mathcal{Y}}^{m}{}_{k}\right)_{(t-\frac{r}{c})} \\[2mm] \tau^{0k} = \tau^{k0} = \tau^{00}n^k. \end{cases}$$

$$(9.2.8)$$

The wave (9.2.2) propagates radially with speed c (fig. 9.4a); the flux Φ of energy radiated in the solid angle $d\Omega$ in direction $\check{n}$ is thus $\Phi = c\tau^{00}r^2 d\Omega$. In the case considered, the only radiation present results from gravitational waves. The power $L_{G/\Sigma}$ lost by gravitational radiation across the sphere Σ

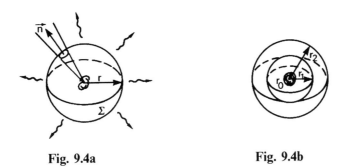

Fig. 9.4a   **Fig. 9.4b**

centred at the origin with very large radius r is the integral of Φ over all directions:

$$L_{G/\Sigma}(t) = \frac{G}{5c^5} \left(\overset{\cdots}{\Psi}{}^{jk} \overset{\cdots}{\Psi}_{jk} \right)_{(t-\frac{r}{c})}. \tag{9.2.9}$$

To derive (9.2.8) we replace $\psi_{\alpha\beta}$ by $h^{(W)}{}_{\alpha\beta}$; in the energy–momentum pseudo-tensor $h^{(W)}{}_{jk}$ is given by (8.5.1), while in line with (8.3.7), $\psi^{\alpha 0} = 0$. In a frame such that $\check{n} = (0, 0, 1)$, we have

$$t^{00} = \tau^{00} = \frac{1}{4\chi c^2} \left(\left(\dot{h}^{(W)}{}_{11} \right)^2 + \left(\dot{h}^{(W)}{}_{22} \right)^2 + 2 \left(\dot{h}^{(W)}{}_{12} \right)^2 \right),$$

$$\{t^{0k}\} = \{\tau^{0k}\} = \tau^{00} \cdot \{0, 0, 1\}.$$

The expression (9.2.2) for $h^{(W)}{}_{jk}$ then leads to (9.2.8). This result is independent of the frame chosen as τ^{00} behaves like a scalar and τ^{0k} like the component of a vector under changes of spatial frame, which is manifestly the behaviour of the right-hand side of (9.2.8); we need only verify the relation in a frame such that $\check{n} = (0, 0, 1)$.

If we consider $h_{\alpha\beta}$ given by (8.3.7) in place of $h^{(W)}{}_{\alpha\beta}$ alone these results remain unchanged: only the quadrupole terms of the radiation contribute to the gravitational energy flux far from the sources.

Consider two spheres of very large radii, r_1 and r_2, centred at the origin. Let L_{G1} and L_{G2} be the values of $L_{G/\Sigma}$ calculated on each of the spheres. Using (9.2.9), we explicitly see that $L_{G1}(t + r_1/c) = L_{G2}(t + r_2/c) \approx L_G(t)$ where $L_G(t)$ represents the energy radiated over a sphere centred at the origin, enclosing the matter distribution, with radius r_0 large enough for (9.2.9) to hold, and small enough that the term r_0/c can be neglected in the argument $t - r/c$ (fig. 9.4b). Within this sphere E_V is equal to E_m up to an additive constant, where E_m is the sum of the Newtonian mechanical energies, the kinetic and potential energies, including the gravitational potential energy. In this case (9.2.7) has the form

$$L_{G/\Sigma}(t) = - \left(\frac{dE_m}{dt} \right)_{(t-\frac{r}{c})}. \tag{9.2.10}$$

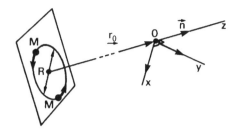

Fig. 9.5

We assume without proof the results leading to (9.2.10). We note only that the slow-motion hypothesis is essential; in particular, if λ is of the order of the wavelengths characteristic of the radiation, the matter must be contained in a sphere of radius R, much smaller than r_0 such that $R \ll r_0 \ll \lambda$.

9.2.2 Order of magnitude of the radiation field at the Earth

We consider a point O, very far from the source, at a distance r_0 in the direction of $\hat{n}$ (fig. 9.5). We choose a frame with origin O, such that $n^1 = n^2 = 0$, $n^3 = 1$. Near O, the gravitational potentials can be written in the following form, which we shall adopt in the rest of this chapter (cf. II § 8.5.3):

♦
$$\begin{cases} h^{(W)}{}_{11} = -h^{(W)}{}_{22} = f_+\left(t - \tfrac{z}{c}\right) \\ h^{(W)}{}_{12} = h^{(W)}{}_{21} = f_\times\left(t - \tfrac{z}{c}\right) \end{cases} \tag{9.2.11}$$

♦
$$\begin{cases} f_+ = -\dfrac{G}{c^4}\dfrac{1}{r_0}\big(\ddot{\ell}^{11} - \ddot{\ell}^{22}\big)_{\left(t-\frac{z}{c}\right)} \\ f_\times = -\dfrac{2G}{c^4}\dfrac{1}{r_0}\big(\ddot{\ell}^{12}\big)_{\left(t-\frac{z}{c}\right)}. \end{cases} \tag{9.2.12}$$

Consider a system formed of two compact objects, of mass M, each in uniform circular motion of radius $R/2$ about their centre of mass in the plane $z = 0$ (fig. 9.5). Using (9.2.9), we determine the value of L_G:

$$L_G = \frac{8G}{5c^5} M^2 R^4 \omega^6 \tag{9.2.13}$$

where ω is the angular velocity of the system.

As the two bodies form a binary system under their Newtonian gravitational attraction, ω and R are not independent: $\omega^2 R^3 = 2GM$. The mechanical energy of the system is then $E_m = -GM^2/2R$.

Assuming that the motions remain quasi-Newtonian with quasi-circular orbits, (9.2.10) gives a differential equation satisfied by the period P:

$$\dot{P}P^{\frac{5}{3}} = -\frac{24}{5}(2\pi)^{\frac{8}{3}}\left(\frac{2GM}{c^3}\right)^{\frac{5}{3}}, \quad P \equiv \frac{2\pi}{\omega}, \quad \dot{P} \equiv \frac{dP}{dt}. \tag{9.2.14}$$

The decrease of E_m through gravitational radiation leads to a decrease in R, an increase in the velocities and thus a decrease of the orbital period. This process is general for systems in near gravitational equilibrium: each time the system loses a joule of total energy, its kinetic energy increases by a joule while its gravitational potential energy decreases by two joules. A similar mechanism makes the stars heat up as they radiate energy. The proof follows from the virial theorem: $2W + U = 0$ where W is the kinetic energy of the system, possibly of thermal origin, and U the gravitational potential energy.

The decrease of orbital period resulting from the loss of energy via gravitational radiation was observed for the first time in the binary system PSR 1913+16 in our Galaxy. This system is composed of two compact objects, in this case neutron stars, of which one is a pulsar with a very stable frequency emitting regular signals with a period near 59 ms. The Doppler effects resulting from the orbital motion are so large and the analysis of the arrival times so precise that relativistic corrections must be taken into account to describe the motion of the system. These include corrections of the type studied in chapter II.3 but also the decrease of the orbital period under the effect of gravitational wave emission. The orbital period P of the binary is close to 28 000 s, or about eight hours; observations give $\dot{P} \approx -2.4 \times 10^{-12}$. General relativity provides a theoretical model with a certain number of free parameters, such as the masses of the two objects, the orbital eccentricity, etc. Adjustment of the theoretical model to the observations of the arrival times of the pulsed signals gives the value of the various parameters; the adjustment is perfect if account is taken of the "braking" of the orbital motion caused by the emission of gravitational waves. Currently the uncertainties are close to 3×10^{-3}. The compact objects both have masses close to 1.4 solar masses; their orbit has eccentricity close to 0.6.

The binary system discovered in 1974 by Russell Hulse and Joseph Taylor using the Arecibo radio telescope has been constantly observed by Taylor. The establishment of the theoretical model allowing a confrontation with observation is a complex operation which had not previously been deeply studied. Observations of the binary pulsar stimulated a number of theoretical investigations over more than a decade, among them work by Thibault Damour at l'Observatoire de Paris. If observation and experiment provide very strong motivations for theoretical research, theoretical studies may animate some experimental developments. The gyroscope experiment discussed in the last section is an example, and the construction of gravitational wave detectors is another.

Solving the differential equation (9.2.14) gives the function $P(t)$. It is interesting to calculate the time for coalescence of the system, i.e. the time after which R will be less than the diameter of a neutron star, which we can take as about ten kilometres, and to compare this with the Hubble time $1/H_0$. We can assume that the current value of $\dot{P}$ is that of the binary pulsar and that M is of order 3×10^{30} kg. This calculation gives a qualitative idea about the existence of binary systems close to coalescence.

To find the order of magnitude h of the gravitational potentials f_+ and $f_\times$ at Earth, we assume that we are situated at O, on the common axis of the two circular orbits. We get

$$f_+ = \frac{2G}{c^4} \frac{1}{r_0} MR^2\omega^2 \cos 2\omega t \quad f_\times = \frac{2G}{c^4} \frac{1}{r_0} MR^2\omega^2 \sin 2\omega t. \quad (9.2.15)$$

With for example $\omega = 3000$ s^{-1}, $M = 2 \times 10^{30}$ kg and $r_0 = 15$ Mpc, we get

$$h \approx (f_+)_{\text{Maximum}} \approx (f_\times)_{\text{Maximum}} \approx 6 \times 10^{-22}. \quad (9.2.16)$$

15 Mpc is the distance to the Virgo cluster, which contains many galaxies; the number of coalescing binaries is therefore possibly large enough for observable phenomena to be plentiful.

> Observation of gravitational waves at Earth provides the period P, and thus ω and $\dot{P}$. The relation (9.2.14) then gives M. As the system is self-gravitating, the value of R can be deduced also. The maximum amplitude of f_+ for example gives r_0, i.e. the distance of the object from us. Binary systems have more independent parameters than the system considered, but we can see from this simple example the importance of a direct observation of gravitational waves. Moreover gravitational waves intersect very little with matter, and they reach us from all directions in space without absorption or distortion, which is a further advantage.

9.2.3 Principle of interferometric detection

To study methods of direct observation of gravitational waves, we must discuss the effects produced in the physical systems used as detectors.

We cannot do this in a purely Newtonian framework as we have to use the relativistic interpretation of the metric $g_{\alpha\beta}$ and that of $h^{(W)}{}_{jk}$ in the quasi-Minkowski coordinate system used:

$$g_{00} = 1, \ g_{0k} = 0, \ g_{jk} = \eta_{jk} + h^{(W)}{}_{jk} \qquad (9.2.17)$$

where $h^{(W)}{}_{jk}$ is given by (9.2.11) and (9.2.12).

> In reality, we must take account of all the sources in the determination of $h_{\alpha\beta}$. In the framework of linearised theory, gravitational perturbations of diverse origins are added in the expression (9.2.17) for the metric. Gravitational radiation may however be distingushed by a temporal analysis of the signal, to the extent that noise of similar frequency is eliminated. This noise may be vibrations, electronic, optical, etc., but also variable Newtonian gravitational fields. As an exercise the reader should perform the calculations following, with a further static perturbation in (9.2.17).

Consider the simplest material system: a dust cloud composed of free particles, without significant gravitational interactions, initially at rest in a Minkowski space-time. Gravitational radiation (9.2.11) reaches the cloud. Writing the equations of motion for test particles (cf. II (2.3.5)) in the metric (9.2.17), we can show that they remain at rest. Of course, this does not mean that there is no physical effect; this simply reflects the choice of coordinate system.

Consider a particle M_0 at the origin and a particle M_1 at $x^k = a^k$. An electromagnetic signal is sent at time t from M_0 towards M_1; it is reflected at M_1 and reaches M_0 at time $t + \tau$ (fig. 9.6). In this case $h_{00} = 0$, and the time τ is the proper time measured in M_0 between the transmission and receipt of the signal.

Assume that $|a^k|$ is small enough that $h_{\alpha\beta}$ remains constant in the space-time region where the phenomena occur.

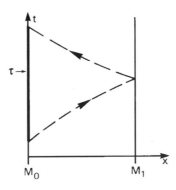

Fig. 9.6

We set $u^k = a^k/L$. To a first approximation $L = \sqrt{\sum a^k a^k}$ is the separation of the two particles. The time τ may be calculated to first order for any $h_{\alpha\beta}$.

Using $ds = 0$, which must hold on the space-time trajectory of the electromagnetic signal, with $h_{00} = 0$, we find

$$\tau = \frac{2L}{c} \left(1 - \frac{1}{2} h_{kj} u^k u^j \right). \qquad (9.2.18)$$

If the metric has the form (9.2.11), taking account of our assumptions, we can set $z = 0$. Thus

$$\tau = \frac{2L}{c} - \frac{L}{c} \sin^2 \theta \left(f_+(t)\cos 2\phi + f_\times(t)\sin 2\phi \right), \quad \text{with} \qquad (9.2.19)$$

$$\{u^k\} = \{\cos \phi \sin \theta,\ \sin \phi \sin \theta,\ \cos \theta\}$$

where t is the time at which the experiment occurs.

$c\tau/2 = L + \delta L$ is the length of the optical path between M_0 and M_1. This length depends on the time t. We note that δL vanishes if $\mathring{u}$ is parallel to the propagation direction of the wave, $\sin \theta = 0$: *the effect is transverse*. Fig. 9.7 gives a polar plot of $1 + \delta L/L$ as a function of ϕ for $\cos \theta = 0$, for sinusoidal gravitational waves with period T.

Assume that the gravitational wave comes from the zenith or nadir. Consider a Michelson interferometer (cf. I section 3.1) whose mirrors are suspended in the horizontal plane $z = 0$ (fig. 9.8). The mirrors are initially at rest. In the expected observational conditions the motion of the mirrors is practically that of a free particle in the horizontal plane; we can then use the previous results.

For a gravitational wave which falls orthogonally on the plane formed by the two arms of the apparatus, such that $f_\times = 0$, the difference in length between the two arms is $2\delta L = f_+(t)L$ when $\phi = 0$; it depends on time. The

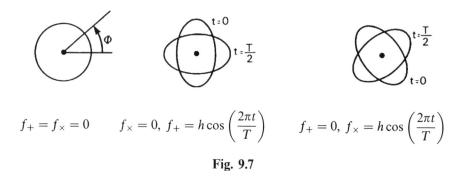

$$f_+ = f_\times = 0 \qquad f_\times = 0,\ f_+ = h\cos\left(\frac{2\pi t}{T}\right) \qquad f_+ = 0,\ f_\times = h\cos\left(\frac{2\pi t}{T}\right)$$

Fig. 9.7

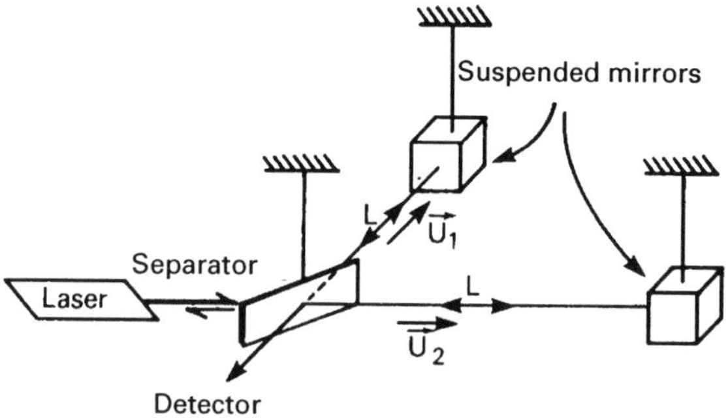

Suspended mirrors

Laser

Separator

Detector

$\vec{U_1}$

$\vec{U_2}$

L

L

L

Fig. 9.8

same holds for the luminous intensity emerging from the apparatus (fig. 9.8). The signal provided by the detector allows us to find $f_+(t)$.

   If the gravitational emission from a binary system is intense, the system shrinks rapidly until coalescence. Such "intense" signals have a short duration. It is thus necessary to have sufficient experimental sensitivity to allow observation of the regions of space which are richest in matter, where these phenomena may be assumed to occur most frequently. A reasonable aim is thus to observe the Virgo cluster at about 15 Mpc. The relation (9.2.16) then gives an estimate of the path difference. It is unlikely that L will exceed a few kilometres: with $L = 3$ km and $2\delta L/L \approx 6 \times 10^{-22}$ for frequencies of order 500 Hz, the variation of the path difference over time is less than 2×10^{-8}Å. Measurement is therefore very difficult, but does not seem beyond the technologies available in 1996 and developed for this purpose in various laboratories: the California and Massachusetts Institutes of Technology, the National Institute for Nuclear Physics in Italy, the Centre National de la Recherche Scientifique in France, and also in Great Britain, Germany and Japan.

To know $h^{(W)}{}_{\alpha\beta}$ completely we have to determine the two angles fixing the direction of the source in the sky, and the axis Oz. We have to find the corresponding functions $f_+(t)$ and $f_\times(t)$. An interferometer provides a single function of time, linearly dependent on f_+ and $f_\times$ as well as the two angles mentioned. Three interferometers are therefore required for complete information without any *a priori* hypothesis. Of course each interferometer is affected by many false signals, but the correlations between the various experiments should allow them to be eliminated.

As an exercise, the reader should explain why the expression (9.2.18) for τ is not gauge invariant, while τ is invariant under coordinate changes.

Show that the interferometer may be arranged so that the measured quantity is first order. Thus verify explicitly that the result of the measurement is gauge invariant.

The expression for τ leads to the relation $\gamma \equiv \dfrac{d^2}{dt^2}(L + \delta L) = Lc^2 R_{0jk0} u^j u^k$.

This is gauge invariant; γ can be intepreted as an acceleration between the particles M_0 and M_1. Such accelerations cause vibrations in solids. Since 1960 Joseph Weber has attempted to exploit this possibility, constructing the first mechanical detectors, metal cylinders weighing more than a tonne, capable of resonating with gravitational waves with frequencies close to a kilohertz. Some of these observations were presented as positive detections. However, these results were not confirmed by any of the many experiments being performed all over the world. Today it is generally agreed that there has as yet been no direct detection of gravitational waves. Several teams, notably in Italy, the USA and Japan, are continuing to develop these detectors and new projects are being considered in Holland and Brazil.

Studies have been undertaken since the beginning of the 1970s to use the idea of interferometric detection of gravitational waves we have discussed. They culminated in 1990 in three projects, LIGO in the USA, GEO in Britain and Germany, and VIRGO in Italy and France. All of these projects aim to construct a Michelson interferometer whose arms, three to four kilometres long, consist of two vacuum tubes with diameters of order a metre (cf. figure p. 185).

The frequency band is limited above by the nature of the astrophysical phenomena, which are generally "slow"; it is limited below by the residual noise of the apparatus and its environment. The usable band thus ranges from a few tens or hundreds of hertz to a few kilohertz.

Currently, in 1996, the construction of LIGO, GEO and VIRGO has started, while the new design of GEO incorporates shorter arms, 600 m long. Shorter prototypes are being developed in Japan.

Appendix A

Lagrange's equations

A.1 Formulation of the problem

Consider a two-dimensional Minkowski spacetime in Minkowski coordinates $\{x, t\}$. Consider two events M and N of this space, with coordinates $\{x_M, t_M\}$ and $\{x_N, t_N\}$, and a curve $\mathscr{C}$ joining the two events. If $\mathscr{C}$ is known we can calculate the proper time τ_{MN}, along $\mathscr{C}$ between M and N. The question we now ask is how to determine $\mathscr{C}$ so that τ_{MN} should be maximal. We thus need a parametric equation for $\mathscr{C}$:

$$t = f(u), \quad x = g(u), \quad \text{with } \{t_M, x_M\} = \{f(0), g(0)\} \text{ and } \{t_N, x_N\} = \{f(1), g(1)\}$$

where the parameter u takes the conventional values 0 and 1 at M and N respectively. The proper time $d\tau$ between two neighbouring events can be written

$$d\tau = \sqrt{dt^2 - \left(\frac{dx}{c}\right)^2} = \sqrt{\left(\frac{df}{du}\right)^2 - \left(\frac{1}{c}\frac{dg}{du}\right)^2} \, du.$$

The functions f and g we seek must therefore give an extremum of the integral τ_{MN}:

$$\tau_{MN} = \int_0^1 L(\dot{f}, \dot{g}) du, \quad \text{with} \quad L(\dot{f}, \dot{g}) = \sqrt{(\dot{f})^2 - \left(\frac{\dot{g}}{c}\right)^2}$$

where we have used the notation $\dfrac{d}{du}(\ ) = (\dot{\ })$.

Thus the problem we have posed can be generalised to the search for an extremal trajectory in an n-dimensional space E_n.

Assume that $2n$ numerical values $\{Q_M{}^k\} = \{Q_M{}^1, Q_M{}^2, ..., Q_M{}^n\}$, and $\{Q_N{}^k\} = \{Q_N{}^1, Q_N{}^2, ..., Q_N{}^n\}$ are given, and represent the coordinates of two points M and N of the space E_n considered.

Let $\{q^k(u)\} = \{q^1(u), q^2(u), ..., q^n(u)\}$ be n *unknown* functions of a parameter u which obey the $2n$ relations

$$q^k(0) = Q^k{}_M \text{ and } q^k(1) = Q^k{}_N \text{ for } k = 1, 2, 3, ..., n. \tag{A.1.1}$$

The curve with equation $x^k = q^k(u)$, where x^k are the coordinates of a point of E_n, passes through M and N for the values 0 and 1 of the parameter.

Let $L(q^k, \dot{q}^j)$ (abbreviated to $L(q, \dot{q})$) be a *given* function of the $2n$ variables $\{q^k, \dot{q}^j\}$. We replace q^k by the function $q^k(u)$, and $\dot{q}^k$ by its derivative with respect to u. We set

$$A[\mathbf{q}] = \int_0^1 L(q, \dot{q})du. \tag{A.1.2}$$

The integral A depends on the chosen functions $q^k(u)$ here represented symbolically by $\mathbf{q}$; we say that A is a **functional** of the $q^k(u)$.

Now consider n arbitrary functions of u, written $\delta q^k(u)$, obeying the relations

$$\delta q^k(0) = \delta q^k(1) = 0. \tag{A.1.3}$$

The curve with equation $x^k = q^k(u) + \varepsilon \delta q^k(u)$, where ε is an arbitrary real number, passes through the points M and N for values 0 and 1 of the parameter. If ε is infinitesimal, expansion to first order gives

$$A[\mathbf{q} + \varepsilon \delta \mathbf{q}] = A[\mathbf{q}] + \varepsilon \delta A. \tag{A.1.4}$$

Finding the extremal trajectory amounts to finding the functions $q^k(u)$ for which A is stationary; that is the particular functions $q_0^k(u)$ for which δA vanishes for any functions $\delta q^k(u)$ satisfying (A.1.3). Posing the problem in this way gives the path along which A is extremal. In the terminology of rational mechanics, the function $L(q, \dot{q})$ is called the **Lagrangian**; A is called the **action**.

A.2 Lagrange's equations

To solve the problem posed above we calculate δA.

$$A[\mathbf{q}] + \varepsilon \delta A = \int_0^1 L(q + \varepsilon \delta q, \dot{q} + \varepsilon \dot{\delta q})du = \int_0^1 [L(q, \dot{q}) + \varepsilon \delta L]du$$

$$\text{with } \delta L = \sum_{k=1}^{k=n} \frac{\partial L}{\partial q^k} \delta q^k + \sum_{k=1}^{k=n} \frac{\partial L}{\partial \dot{q}^k} \delta \dot{q}^k. \tag{A.2.1}$$

To calculate δL we have made a first-order Taylor expansion of L in ε. The partial derivatives of L appearing in (A.2.1) are to be understood as derivatives with respect to the $2n$ independent variables q^k and $\dot{q}^j$.

Identifying the terms of first order in ε in (A.2.1), we find

$$\delta A = \int_0^1 \left(\sum_{k=1}^{k=n} \frac{\partial L}{\partial q^k} \delta q^k + \sum_{k=1}^{k=n} \frac{\partial L}{\partial \dot{q}^k} \delta \dot{q}^k \right) du,$$

and integrating by parts

$$\delta A = \int_0^1 \left(\sum_{k=1}^{k=n} \frac{\partial L}{\partial q^k} \delta q^k - \sum_{k=1}^{k=n} \frac{d}{du} \left(\frac{\partial L}{\partial \dot{q}^k} \right) \delta q^k \right) du + \left[\sum_{k=1}^{k=n} \frac{\partial L}{\partial \dot{q}^k} \delta q^k \right]_0^1.$$

We note that use of (A.1.3) shows that the last term vanishes. Thus δA vanishes *whatever the functions* δq^k in $]0, 1[$ iff

♦
$$\frac{\partial L}{\partial q^k} - \frac{d}{du} \left(\frac{\partial L}{\partial \dot{q}^k} \right) = 0, \text{ for } k = 1, 2, \ldots, n. \qquad \text{(A.2.2)}$$

These relations are known as **Lagrange's equations** in rational mechanics. They are necessarily satisfied by the functions $q_0{}^k(u)$ we seek. We note that Lagrange's equations are local differential equations and that they are independent of the starting points M and N. Similarly from the calculations above we can see that the equations are unchanged if the parameter values at M and N differ from 0 and 1.

The Lagrange equations derived from an arbitrary function L are not necessarily soluble: for example $L = \sum_k (q^k + \dot{q}^k)$.

Several Lagrangians may lead to the same equations: for example $L(q, \dot{q})$, $C \cdot L$, $L + C$ and $L + \sum_k \dot{q}^k \frac{\partial F(q)}{\partial q^k}$, where C is a constant and F an arbitrary function of q^k.

Also for $L = \sum_k \dot{q}^k \frac{\partial F(q)}{\partial q^k}$, Lagrange's "equations" are always satisfied regardless of the functions $q_0{}^k(u)$.

A.3 Motion of a free particle in special relativity

To determine the motion of a free particle in *special relativity* we identify the proper time with the action. In a Minkowski system we set:

$$L = \sqrt{ \left(\frac{dt}{du} \right)^2 - \left(\frac{1}{c} \frac{dx}{du} \right)^2 - \left(\frac{1}{c} \frac{dy}{du} \right)^2 - \left(\frac{1}{c} \frac{dz}{du} \right)^2 }, \text{ with}$$

$$\{x, y, z, t\} = \{q^1, q^2, q^3, q^4\}.$$

Lagrange's equations can be written

$$\frac{d}{du} \left(\frac{1}{L} \frac{dx}{du} \right) = \frac{d}{du} \left(\frac{1}{L} \frac{dy}{du} \right) = \frac{d}{du} \left(\frac{1}{L} \frac{dz}{du} \right) = \frac{d}{du} \left(\frac{1}{L} \frac{dt}{du} \right) = 0.$$

They have the first integral

$$\frac{1}{L} \frac{dx}{du} = A_1, \quad \frac{1}{L} \frac{dy}{du} = A_2, \quad \frac{1}{L} \frac{dz}{du} = A_3, \quad \frac{1}{L} \frac{dt}{du} = A_4, \text{ where } A_k = \text{constant}.$$

We thus easily find x, y and z as functions of time; for example

$$\frac{dx}{dt} = \frac{A_1}{A_4} \equiv V^1 \Leftrightarrow x = V^1 t + a^1,$$

where V^1 and a^1 are arbitrary constants. The functions $y(t)$ and $z(t)$ can be found similarly. The motion of a free particle is therefore rectilinear in the coordinate system used.

A.4 Particular Lagrangians

The equations governing many dynamical systems are Lagrange equations for simple choices of Lagrangian. For example in *Newtonian mechanics*, the motion of a purely passive particle of mass m in an external field such that the potential energy of the system is $U(x, y, z)$ depends only on the position of the particle. In this case the absolute time can be used as a parameter to describe the motion in space. The Lagrangian L has the form

$$L = \frac{1}{2} m \vec{V}^2 - U(x, y, z), \text{ with}$$

$$\{x, y, z\} = \{q^1, q^2, q^3\}, \ \vec{V} = \{\dot{x}, \dot{y}, \dot{z}\} \text{ and } (\dot{\ }) = \frac{d}{dt}(\ ).$$

Lagrange's equations are

$$\frac{dm\vec{V}}{dt} = -\overrightarrow{\text{grad}[U]} \Leftrightarrow \vec{F} = m\vec{\gamma}, \text{ with } \vec{F} \equiv -\overrightarrow{\text{grad}[U]}$$

where $\vec{\gamma}$ is the particle's acceleration.

Multiplying each side of this equation by $\vec{V}$, we find

$$\vec{V} \cdot \frac{d(m\vec{V})}{dt} = \frac{d}{dt}\left[\frac{1}{2}m\vec{V}^2\right] = -\vec{V} \cdot \overrightarrow{\text{grad}[U]} = -\frac{d}{dt}[U]$$

$$\Rightarrow \frac{1}{2}m\vec{V}^2 + U = \text{ constant.}$$

The latter relation expresses the conservation of mechanical energy: kinetic + potential. This conservation can be substituted for one of the Lagrange equations.

It is sometimes convenient to use a non-Cartesian coordinate system, such as spherical polars for example. The corresponding Lagrangian can be found by expressing the kinetic energy as a function of the new coordinates r, θ, ϕ: $x = r \sin \theta \cos \phi$, $y = r \sin \theta \sin \phi$, $z = r \cos \theta$:

$$\vec{V}^2 = \frac{(d\ell)^2}{(dt)^2} = \frac{(dr^2 + r^2 \sin^2 \theta \ d\phi^2 + r^2 d\theta^2)}{dt^2}$$

$$L = \frac{1}{2}m(\dot{r}^2 + r^2 \sin^2 \theta \ \dot{\phi}^2 + r^2 \dot{\theta}^2) - U$$

where U is regarded as a function of r, θ, ϕ. We easily verify that the equations of motion in spherical polars can be found from the preceding Lagrangian with $\{q^1, q^2, q^3\} = \{r, \theta, \phi\}$ and $t = u$.

A particularly important case in relativity involves Lagrangians which are quadratic in the derivatives.

Consider a Lagrangian of the form

$$L = \sum_{j=1}^{n} \sum_{k=1}^{n} G_{jk}(q) \dot{q}^j \dot{q}^k \tag{A.4.1}$$

where $G_{jk}(q)$ is a function only of the q^k.

Setting $G_{jk} = \frac{1}{2}(G_{jk} + G_{kj}) + \frac{1}{2}(G_{jk} - G_{kj})$ we note that the second, antisymmetric term of the decomposition adds a zero contribution to L. We can thus assume that $G_{jk} = G_{kj}$, without loss of generality.

Lagrange's equations are

$$\sum_j \sum_k G_{jk,m}(q) \dot{q}^j \dot{q}^k - \frac{d}{du} \sum_j 2G_{jm}(q) \dot{q}^j = 0, \text{ with}$$

$$G_{jk,m}(q) \equiv \frac{\partial G_{jk}}{\partial q^m}. \tag{A.4.2}$$

Expanding (A.4.2) we get

♦
$$\sum_{(j,k)} G_{jk,m}(q) \dot{q}^j \dot{q}^k - \sum_{(j,k)} 2G_{jm,k}(q) \dot{q}^j \dot{q}^k - \sum_j 2G_{jm}(q) \ddot{q}^j = 0. \tag{A.4.3}$$

Multiplying each side by q^m and summing over all m from 1 to n, we find the relation

$$\sum_{(j,k,m)} G_{jk,m} \dot{q}^j \dot{q}^k \dot{q}^m + \sum_{(j,m)} 2G_{jm} \ddot{q}^j \dot{q}^m \equiv \frac{dL}{du} = 0. \tag{A.4.4}$$

It is often convenient to replace one of the Lagrange equations by the first integral $L = $ constant.

A.5 Change of parameter

The solution of the physical problem is the curve $\mathscr{C}$ of E_n making the action stationary; the parametrisation of this curve is simply a question of convenience, without any physical meaning. Here we study changes of parametrisation in the physical problem described by the quadratic Lagrangian above, (A.4.1).

We introduce a new parameter v which is a function of u: $v = F(u)$, and we assume that F is a monotonic function, which can be inverted as $u = F^{-1}(v)$. The coordinates x^k are functions of v: $x^k = \ell^k(v) = q^k(u) = \ell^k[F(u)]$. The equations satisfied by ℓ^k can be found from equation (A.4.3).

$$0 = \sum_{(j,k)} G_{jk,m}(\ell) \cdot (\ell^j)'(\ell^k)'f^2(v) - \sum_{(j,k)} 2G_{jm,k}(\ell) \cdot (\ell^j)'(\ell^k)'f^2(v)$$
$$- \sum_j 2G_{jm}(\ell) \cdot (\ell^j)''f^2(v) - \sum_j 2G_{jm}(\ell) \cdot (\ell^j)'f(v)f'(v) \qquad (A.5.1)$$

where we have set $\ell^k = \ell^k(v)$, $f(v) = \left(\dfrac{dF}{du}\right)_{u=F^{-1}(v)}$ and $(\ )' \equiv \dfrac{d(\ )}{dv}$.

In general $f' \neq 0$, and equations (A.5.1) satisfied by ℓ^k differ from equations (A.4.3), satisfied by q^k; however, if u and v are related by an affine transformation, $v = au + b$, the two sets of equations are formally identical. The equations satisfied by $\ell^k(v)$ are then deducible from the Lagrangian $\Sigma G_{jk}(\ell) \cdot (\ell^j)' \cdot (\ell^k)'$ which is the Lagrangian (A.4.1) up to notation. Thus giving a Lagrangian determines the parametrisation up to an affine transformation.

This freedom can be used to impose the value of the parameter at M and N *a priori*. If we prefer, we can using (A.4.4) impose a parametrisation such that along $\mathscr{C}$, $|L| = 1$ or $L = 0$.

For a given Lagrangian the freedom to change the parametrisation by an affine transformation is usually *insufficient* to give an immediate physical sense to the parameter. For example, if we set $u = q^1$ *in the Lagrangian* the solutions of the Lagrange equations for $q^2, q^3, ..., q^n$ do not represent the curve $\mathscr{C}$ in the form $q^k = q^k(q^1)$.

A.6 Free-fall motion of a test particle

In this section we shall use the Einstein summation convention for repeated indices, one raised and one lowered. The dynamical variables q^k are here the coordinates x^α of the space–time points; Greek indices take the values 0, 1, 2 and 3.

According to the physical reasoning given in the second chapter of part II, the trajectory of a freely falling point mass is a curve $\mathscr{C}_0$ on which the proper time τ_{AB} between any two points A and B is stationary:

$$\tau_{AB} = \frac{1}{c} \int_{u_A}^{u_B} \sqrt{g_{\alpha\beta}(x)\dot{x}^\alpha \dot{x}^\beta}\, du$$

where the metric $g_{\alpha\beta}$ is a function only of the space-time coordinates x^α of the particle.

There is no difference in assuming that the proper time τ_{AB} is stationary, or that the proper length $s_{AB} = c\tau_{AB}$ is stationary. The curve $\mathscr{C}_0$ is thus determined by the equations derived from the Lagrangian

$$L = \sqrt{g_{\alpha\beta}(x)\dot{x}^\alpha \dot{x}^\beta}. \qquad (A.6.1)$$

Lagrange's equations are

$$\frac{1}{2L}g_{\alpha\beta,\sigma}\dot{x}^{\alpha}\dot{x}^{\beta} - \frac{d}{du}\left(\frac{1}{L}g_{\alpha\sigma}\dot{x}^{\alpha}\right) = 0 \text{ with } (\ )_{,\mu} \equiv \partial_{\mu}(\ ) \equiv \frac{\partial(\ )}{\partial x^{\mu}}. \quad (A.6.2)$$

Under the same conditions as in the preceding section, we can make a change of parametrisation $v = F(u)$. The coordinates are functions $y^{\alpha}(v)$ such that $y^{\alpha}(v) = x^{\alpha}(u) = y^{\alpha}[F(u)] = x^{\alpha}[F^{-1}(v)]$. We easily verify that the equations satisfied by $y^{\alpha}(v)$ are equations (A.6.2) up to notation (v and derivatives with respect to it replace u and the corresponding derivatives while y^{α} replaces x^{α}). They can therefore be deduced from the Lagrangian (A.6.1) by making the same substitutions at the outset. *This type of Lagrangian therefore does not restrict the parametrisation.* Assume that we have solved Lagrange's equations. We can choose $u = as + b$, where a and b are arbitrary constants with $a \neq 0$ and s the proper length measured along $\mathscr{C}_0$ from an arbitrary origin. Such a parameter is called an **affine parameter**. In using this parameter we impose at each point of $\mathscr{C}_0$ the relation $L|a| = 1$ without changing the expression (A.6.1) for the Lagrangian. Lagrange's equations are then

$$g_{\sigma\alpha}\ddot{x}^{\alpha} + \frac{1}{2}(2g_{\sigma\alpha,\beta} - g_{\alpha\beta,\sigma})\dot{x}^{\alpha}\dot{x}^{\beta} = 0. \quad (A.6.3)$$

The corresponding curves $\mathscr{C}_0$ are called **geodesics** of the space–time.

We note that the Lagrangian introduced in § A.3 has the form (A.6.1). We can therefore choose *a priori* the time t as parameter. Lagrange's equations are not changed if we multiply the Lagrangian by the constant $-mc^2$ where m is the mass of the particle considered. Under these conditions we have

$$L = -mc^2\sqrt{1 - \frac{\vec{V}^2}{c^2}}$$

where $\vec{V}$ is the particle's velocity with respect to the Minkowski system used. The Lagrangian is therefore measured in joules and the action in joule seconds (J s). Expansion to lowest order in $|V/c|$ leads to the Newtonian Lagrangian for a free particle up to an additive constant $-mc^2$; this constant has no effect on the Lagrange equations.

Equations (A.6.3) can also be obtained starting from the quadratic Lagrangian $\mathscr{L}$:

$$\mathscr{L} = g_{\alpha\beta}(x)\,\dot{x}^{\alpha}\dot{x}^{\beta}. \quad (A.6.4)$$

The Lagrange equations from the quadratic Lagrangian thus give the equations of $\mathscr{C}_0$ with affine parameter.

There are several advantages in using the quadratic Lagrangian. We can easily extend the idea of a point mass in free fall to massless particles; in this case $\mathscr{L} = 0$ on $\mathscr{C}_0$. Unlike the Lagrangian L, the quadratic Lagrangian does not require the curves considered to be timelike in order to be defined. The arbitrariness in the choice of affine parameter allows us to set *a priori* the condition $\mathscr{L} = 1$ for timelike curves, with $g_{\alpha\beta}\dot{x}^{\alpha}\dot{x}^{\beta} > 0$, and $\mathscr{L} = -1$ for spacelike curves, with $g_{\alpha\beta}\dot{x}^{\alpha}\dot{x}^{\beta} < 0$. These relations, given here as conditions on the parametrisation, are the first integrals mentioned earlier (§ A.4).

Appendix B

Geometric framework of general relativity

We have seen in the second chapter of part II how Einstein's equivalence principle leads to the introduction of a "metric" $g_{\alpha\beta}(x)$ transforming according to (II 2.4.2) under coordinate changes. The equations of motion for free fall of a point particle were found in an analogous form to that in special relativity in an arbitrary coordinate system. More generally the set of relations in local coordinates summarised in table 7.1 of part I remain in force for the class of theories considered. To explain the mathematical basis of this analogy we shall here describe the geometrical framework in which theories with a metric connection are developed. This is the same as the geometrical framework of gauge theories. We shall give a treatment allowing this extension although limiting ourselves to specific aspects of gravity theories.

B.1 The notion of a fibre

Events are the points of a four-dimensional space–time manifold V_4 which reduces to the space M_4 in special relativity. We denote the coordinates as x^μ, or x for brevity, with $\mu = 0$, 1, 2 or 3. At each point of V_4 we introduce an N-dimensional vector space $F(x)$, which can be constructed over the field K of complex numbers or the reals. We limit ourselves here to the case where K is the real numbers. The local physical state at x of the physical system considered is described by a vector of $F(x)$. The space $F(x)$ generalises the space T_4 introduced in special relativity, with however the essential difference that T_4 is unique, while $F(x)$ and $F(x')$ are distinct spaces at different points, and the dimension of $F(x)$ is for the moment arbitrary. A structure of this kind constitutes a **fibre bundle**: V_4 is the **base** and F is the **fibre**; the point with coordinates $\{x^\mu\}$ is the "projection" of $F(x)$ onto the base. Vectors of $F(x)$ will be denoted $\vec{f}(x)$. We shall use a basis $\{\vec{e}_a(x)\}$ of $F(x)$ which we assume known: $a = 1, 2, \ldots, N$. If there is no danger of ambiguity we shall omit the arguments x.

B.2 Linear connection

To construct a verifiable physical theory we need a way of representing "identical" physical states at different places and times, so that experiments can therefore be regarded as repeatable. This implies that we can compare physical states at two neighbouring points x and $x + dx$.

Given a vector $\vec{f}_{(x+dx)}$ belonging to $F(x + dx)$, we write $\overrightarrow{\text{Im}}_{(x,x+dx)}[\vec{f}_{(x+dx)}]$ for the vector of $F(x)$ which represents the "identical" physical state at x to that represented by $\vec{f}_{(x+dx)}$ at $x + dx$. In the following we simplify the notation to $\overrightarrow{\text{Im}}$, for the "image" in place of $\overrightarrow{\text{Im}}_{(x,x+dx)}$. We demand linearity of the operation $\overrightarrow{\text{Im}}$:

$$\overrightarrow{\text{Im}}\left[\mu \vec{f}_{1_{(x+dx)}} + \vec{f}_{2_{(x+dx)}}\right] = \mu \overrightarrow{\text{Im}}\left[\vec{f}_{1_{(x+dx)}}\right] + \overrightarrow{\text{Im}}\left[\vec{f}_{2_{(x+dx)}}\right] \in F(x) \qquad \text{(B.2.1)}$$

where μ belongs to K and where $\vec{f}_{1_{(x+dx)}}$ and $\vec{f}_{2_{(x+dx)}}$ are arbitrary vectors of $F(x + dx)$.

The correspondence $\overrightarrow{\text{Im}}_{(x,x+dx)}$ is completely defined once the image at x of basis vectors of $F(x + dx)$ is given:

$$\overrightarrow{\text{Im}}\left[\vec{e}_a(x + dx)\right] = \vec{e}_a(x) + \Gamma^b{}_{a\mu}(x)dx^\mu \vec{e}_b(x). \qquad \text{(B.2.2)}$$

The coefficients $\Gamma^b{}_{a\mu}(x)$ are the **linear connection** coefficients defining the operation $\overrightarrow{\text{Im}}$ to first order in dx.

We set $\vec{f}(x + dx) \equiv f^a \vec{e}_a(x + dx)$ and $\overrightarrow{\text{Im}}[\vec{f}(x + dx)] \equiv f'^a \vec{e}_a(x)$.

From (B.2.1) and (B.2.2) we get

$$f'^a = f^a + \Gamma^a{}_{b\mu} f^b dx^\mu. \qquad \text{(B.2.3)}$$

B.3 Absolute differential and parallel transport

Consider a vector field $\vec{f}(x)$ specified at each x by its components $f^a(x)$ in the basis $\{\vec{e}_a(x)\}$, chosen at each point x. Consider two neighbouring points and the corresponding vectors $\vec{f}(x) \in F(x)$, $\vec{f}(x + dx) \in F(x + dx)$. We define the **absolute differential** of the two vectors, $\nabla \vec{f}$ (fig. B.1):

$$\nabla \vec{f} \equiv \overrightarrow{\text{Im}}[\vec{f}(x + dx)] - \vec{f}(x) \equiv (\nabla f^a)\vec{e}_a(x) \in F(x). \qquad \text{(B.3.1)}$$

Using the relations (B.2.3) we find to first order in dx^μ

$$\nabla f^a = df^a + \Gamma^a{}_{b\mu} f^b dx^\mu; \quad df^a \equiv \partial_\mu f^a dx^\mu; \quad \partial_\mu \equiv \frac{\partial}{\partial x^\mu}. \qquad \text{(B.3.2)}$$

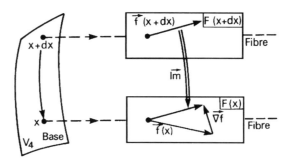

Fig. B.1

For preference we write

$$\nabla \vec{f} = \nabla f^a \vec{e}_a = \nabla_\mu f^a dx^\mu \vec{e}_a; \quad \nabla_\mu f^a \equiv \partial_\mu f^a + \Gamma^a{}_{b\mu} f^b \qquad \text{(B.3.3)}$$

where $\nabla_\mu f^a$ is called the **covariant derivative** of the vector field at the point with coordinates x.

For the basis vectors we get

$$\nabla \vec{e}_b = \Gamma^a{}_{b\mu} dx^\mu \vec{e}_a. \qquad \text{(B.3.4)}$$

When we compare two isolated vectors defined at neighbouring points which do not belong to a vector field, the relation (B.3.1) defines the **absolute difference** of the two vectors, to first order in dx^μ.

Consider a vector at a given point A of V_4 and a curve of V_4 passing through this point. We displace this vector along the curve in such a way that its absolute differential, defined by the relation (B.3.1), remains zero. This is called **parallel transport**. The physical state described by this vector remains "identical to itself" along the curve. If the curve is parametrised by s, according to (B.3.2), the equations satisfied by the components $f^a(s)$ of the vector can be written as

$$\frac{df^a}{ds} + \Gamma^a{}_{b\mu} f^b \frac{dx^\mu}{ds} = 0. \qquad \text{(B.3.5)}$$

In this expression, assuming the curve given by functions $x^\mu(s)$, the various quantities f^a, x^μ, $\Gamma^a{}_{b\mu}(x)$ are functions of s. The initial conditions, i.e. the value of f^a at A, then completely determine the vector along the curve.

Now consider a scalar field, i.e. a function $g(x)$, defined on V_4 and taking values in K. Such a function might for example represent the proportions of the various constituents of a chemical mixture which varies from point to point. The physical interpretation of the absolute differential identifies it with the ordinary differential. The covariant derivative of a scalar is equal to its partial derivative.

$$\nabla g \equiv dg; \quad \nabla_\mu g(x) \equiv \partial_\mu g(x). \qquad \text{(B.3.6)}$$

B.4 Tangent space

Consider the scalar functions $f(x)$ defined on V_4. We call a **derivation** at the point x_0, any operator $\mathbf{V}_0$, which maps sufficiently regular functions f into the element $\mathbf{V}_0[f]$ of K, such that

$$\begin{cases} \mathbf{V}_0[\mu f_1 + f_2] = \mu \mathbf{V}_0[f_1] + \mathbf{V}_0[f_2] \\ \mathbf{V}_0[f_1 f_2] = \mathbf{V}_0[f_1]f_2(x_0) + f_1(x_0)\mathbf{V}_0[f_2]. \end{cases} \tag{B.4.1}$$

The set of operators $\mathbf{V}_0$ forms a four-dimensional vector space over K. This space, denoted $T(x_0)$, is distinct from the space $T(x_1)$ constructed similarly at the point x_1. These spaces have the same structure however. A basis for $T(x_0)$ is given by the four partial derivatives ∂_μ at the point x_0. To emphasise the vector space structure the derivatives will be written $\vec{\partial}_\mu$ and the elements of $T(x)$ written as $\vec{V}_{(x)}$, or $\vec{V}(x)$ according as we are dealing with a vector defined in $T(x)$ for given x or a vector field defined in $T(x)$ for all x. The space $T(x)$ is called the **tangent space** at x; its elements $\vec{V}$ are called **tangent vectors**.

$$\vec{V}(x) = V^\mu \cdot \vec{\partial}_\mu. \tag{B.4.2}$$

The basis $\{\vec{\partial}_\mu\}$ is called the **natural basis** associated with the coordinate system. If necessary we could supplement the notation to show that $\vec{\partial}_\mu$ is considered at the point x, for example $^{(x)}\vec{\partial}_\mu$. As there is no ambiguity we shall omit this in the following.

The tangent space plays an important role in relativity. It provides the representation of the electromagnetic four-potential; it contains the tangent vectors to the worldlines of particles, such as the four-velocity for example. The energy–momentum tensor of a fluid is an element of the tensor product of the tangent space with itself, and so on.

As an exercise the reader may show that ∂_μ is a derivation.

To give $T(x_0)$ the structure of a vector space we define the sum of two derivations and multiplication of a derivation by an element of K in the following way for any derivations $\mathbf{V}_0$ and $\mathbf{W}_0$ of $T(x_0)$ and any element, μ, of K:

$(\mathbf{V}_0 + \mu \mathbf{W}_0)[g] \equiv \mathbf{V}_0[g] + \mu \mathbf{W}_0[g]$, $\forall g(x)$ sufficiently regular in the neighbourhood of x_0.

We easily verify that $(\mathbf{V}_0 + \mu \mathbf{W}_0)$ defined in this way is a derivation and that $T(x_0)$ is a vector space with these rules.

To prove (B.4.2) we first consider the effect of $\mathbf{V}_0$ on a constant function μ. From (B.4.1) we get $0 = \mathbf{V}_0[0 \cdot 0] = \mathbf{V}_0[0]$ and $\mathbf{V}_0[\mu f] = \mu \mathbf{V}_0[f] = \mathbf{V}_0[\mu] \cdot f(x_0) + \mu \mathbf{V}_0[f]$, implying $\mathbf{V}_0[\mu] = 0$. We assume that in a neighbourhood of x_0 $f(x)$ can be expanded as

$$f(x) = f(x_0) + (x^\mu - x_0{}^\mu)\{\partial_\mu f\}_0 - (x^\mu - x_0{}^\mu) \cdot H_\mu(x) \text{ where } H_\mu(x_0) = 0,$$

which for example holds for functions which can be Taylor-expanded in the neighbourhood of x_0. Using relations (B.4.1) and the previous result, we get $\mathbf{V}_0[f] = V_0{}^\mu \cdot \{\partial_\mu f\}_0$ with $V_0{}^\mu = \mathbf{V}_0[x^\mu]$ demonstrating (B.4.2). Hence the vector

$\mathbf{V}_0$, which we write $\vec{V}_0$, is completely determined by its four components V_0^μ. These can be chosen arbitrarily in K.

B.5 Connection associated with the tensor product of two spaces

Consider at each point x two vector spaces $F(x)$ and $\Phi(x)$ with dimensions N and M. The bases of F and Φ are written $\{\vec{e}_a\}$ and $\{\vec{\varepsilon}_k\}$ with $a = 1, 2, \ldots, N$ and $k = 1, 2, \ldots, M$. The tensor product $F(x) \otimes \Phi(x)$ is written $S(x)$. The basis of $S(x)$ we consider is $\vec{\sigma}_{(ak)} = \vec{e}_a \otimes \vec{\varepsilon}_k$. To retain the physical meaning of parallel transport we define the correspondence $\overrightarrow{\text{Im}}$ between $S(x + dx)$ and $S(x)$ in the following way:

$$\overrightarrow{\text{Im}}[\vec{f} \otimes \vec{\phi}] = \overrightarrow{\text{Im}}[\vec{f}] \otimes \overrightarrow{\text{Im}}[\vec{\phi}] \qquad (B.5.1)$$

where $\vec{f}$ and $\vec{\phi}$ are two arbitrary vectors of $F(x+dx)$ and $\Phi(x+dx)$ respectively. This relation leads to the following expression for the connection of $S(x)$:

$$\Gamma^{(ak)}{}_{(bj)\mu} = \delta^a{}_b \Gamma^k{}_{j\mu} + \delta^k{}_j \Gamma^a{}_{b\mu} \qquad (B.5.2)$$

where $\Gamma^a{}_{b\mu}$ and $\Gamma^k{}_{j\mu}$ are the connections of $F(x)$ and $\Phi(x)$ respectively, while $\delta^a{}_b$ and $\delta^k{}_j$ are Kronecker symbols in these spaces.

We deduce the expression for the covariant derivative of a tensor field $T^{ak}(x) \cdot \vec{\sigma}_{(ak)}(x)$ belonging to $S(x)$:

$$\nabla_\mu T^{ak} = \partial_\mu T^{ak} + \Gamma^a{}_{b\mu} T^{bk} + \Gamma^k{}_{j\mu} T^{aj}. \qquad (B.5.3)$$

In the particular case $T^{ak} = f^a \phi^k$, we get

$$\nabla_\mu(f^a \phi^k) = (\nabla_\mu f^a)\phi^k + f^a \nabla_\mu(\phi^k). \qquad (B.5.4)$$

This is known as **Leibniz's rule**.

B.6 Change of frame

A change of frame is a change of basis in the fibre F; it is characterised by invertible matrices $\ell^b{}_{a'}(x)$ whose inverse matrices are written $\ell^{a'}{}_b(x)$:

$$\ell^b{}_{a'}\ell^{a'}{}_c = \delta^b{}_c \quad \text{and} \quad \ell^{a'}{}_b\ell^b{}_{c'} = \delta^{a'}{}_{c'}. \qquad (B.6.1)$$

The correspondence between the new basis $\{\vec{e}_{a'}\}$ and the old basis $\{\vec{e}_a\}$ is determined at each point x by the relations

$$\vec{e}_{a'} = \ell^b{}_{a'}\vec{e}_b; \quad \vec{e}_a = \ell^{b'}{}_a\vec{e}_{b'}. \qquad (B.6.2)$$

In a change of frame the components of a vector, the connection coefficients, and the expression for the covariant derivative are modified in the following ways, which we leave to the reader to verify:

$$
\begin{cases}
\vec{f}(x) \equiv f^a(x)\vec{e}_a(x) \equiv f^{a'}(x)\vec{e}_{a'}(x) \\
f^{a'}(x) = \ell^{a'}{}_b(x)f^b(x), \quad f^a(x) = \ell^a{}_{b'}(x)f^{b'}(x) \\
\Gamma^{a'}{}_{b'\mu} = \ell^{a'}{}_c\ell^d{}_{b'}\Gamma^c{}_{d\mu} + \ell^{a'}{}_c\partial_\mu\ell^c{}_{b'}, \quad \nabla_\mu f^{a'} = \ell^{a'}{}_b\nabla_\mu f^b.
\end{cases}
\tag{B.6.3}
$$

At each point we can introduce the dual $F^*(x)$ of $F(x)$, the dual basis of $\{\vec{e}_a(x)\}$, written $\{e^a(x)\}$, and the corresponding connection $\Gamma^{*a}{}_{b\mu}$. This defines parallel transport in $F^*(x)$, i.e. the operation $\overrightarrow{\mathrm{Im}}^*{}_{[x,x+dx]}$:

$$
e^a[\vec{e}_b] \equiv \delta^a{}_b; \quad \overrightarrow{\mathrm{Im}}^*[e^a] \equiv e^a + \Gamma^{*a}{}_{b\mu}dx^\mu e^b; \quad \nabla e^a \equiv \Gamma^{*a}{}_{b\mu}dx^\mu e^b.
\tag{B.6.4}
$$

By convention, F^* has a frame given by the dual basis of F, so a change of frame affects the bases of both spaces simultaneously.

The quantities $\ell^a{}_{b'}$ are defined by (B.6.2). In a change of frame we get

$$
\begin{cases}
e^{a'} = \ell^{a'}{}_b e^b, \quad e^a = \ell^a{}_{b'}e^{b'}, \\
f(x) = f_a(x)e^a(x) = f_{a'}(x)e^{a'}(x), \quad f_a(x) = \ell^{b'}{}_a f_{b'}(x) \\
\Gamma^{*a'}{}_{b'\mu} = \ell^{a'}{}_c\ell^d{}_{b'}\Gamma^{*c}{}_{d\mu} - \ell^{a'}{}_c\partial_\mu\ell^c{}_{b'}, \quad \nabla_\mu f_{a'} = \ell^b{}_{a'}\nabla_\mu f_b.
\end{cases}
\tag{B.6.5}
$$

Quite generally, the transformation rules for components of tensors of $\otimes^n F(x) \otimes^m F^*(x)$ are similar to those we introduced in chapter 5 of part I; for example

$$
T^{a'}{}_{b'} = \ell^{a'}{}_c\ell^d{}_{b'}T^c{}_d.
\tag{B.6.6}
$$

These rules characterise a tensor.

B.7 Coordinate changes

A change of coordinates is defined by the four functions

$$
x^{\mu'} = X^{\mu'}(x^\alpha).
\tag{B.7.1}
$$

The system (B.7.1) is assumed invertible. As in special relativity (chapter 7 of part I), we introduce matrices $a^{\alpha'}{}_\beta$ and $a^\mu{}_{\beta'}$, inverse to each other:

$$
a^{\alpha'}{}_\beta(x) \equiv \frac{\partial X^{\alpha'}}{\partial x^\beta}; \quad a^{\alpha'}{}_{\beta'}a^{\beta'}{}_\sigma = \delta^{\alpha'}{}_\sigma.
\tag{B.7.2}
$$

Consider a point E with coordinates $\{x^\alpha\}$ and $\{x^{\alpha'}\}$ in each of the two systems. In a change of coordinates at any point E, the natural basis of the tangent space, the components of a vector, $\vec{V}$, the connection coefficients and the covariant derivative obey the following transformations:

$$
\begin{cases}
\vec{V}_E \equiv V^\mu(x)\vec{\partial}_\mu \equiv V^{\mu'}(x')\vec{\partial}_{\mu'} \\
\vec{\partial}_{\mu'} = a^\alpha{}_{\mu'}\vec{\partial}_\alpha, \quad V^{\mu'}(x') = a^{\mu'}{}_\beta(x)V^\beta(x) \\
\Gamma^a{}_{b\mu'} = a^\alpha{}_{\mu'}\Gamma^a{}_{b\alpha} \quad \nabla_{\mu'}f^a = a^\alpha{}_{\mu'}\nabla_\alpha f^a.
\end{cases}
\tag{B.7.3}
$$

These transformation rules show that $\nabla_\mu f^a$ is the component of a tensor of $T^*(x) \otimes F(x)$, where T^* is the dual of T. The dual basis of $\vec{\partial}_\mu$ is written ∂^μ:

$$\begin{cases} \partial^\mu \left[\vec{\partial}_\alpha\right] \equiv \delta^\mu{}_\alpha \\ \nabla \vec{f} \equiv \nabla_\mu f^a \partial^\mu \otimes \vec{e}_a \in T^*(x) \otimes F(x). \end{cases} \tag{B.7.4}$$

When F and T are two different spaces, changes of basis in F and T are independent. When F is the same as T, T^* or the tensor product of these spaces in any number, we usually use in T the natural basis associated with the coordinate system and in T^* the dual basis; thus by convention a change of coordinates is always associated with a change of basis. In this case the transformations of the connection coefficients become

♦ $$\Gamma^{\alpha'}{}_{\beta'\mu'} = a^\tau{}_{\mu'} \left(a^{\alpha'}{}_\sigma a^\rho{}_{\beta'} \Gamma^\sigma{}_{\rho\tau} + a^{\alpha'}{}_\sigma \partial_\tau a^\sigma{}_{\beta'} \right). \tag{B.7.5}$$

The connection coefficients do not constitute the components of a tensor in $\otimes^n T \otimes^m T^*$; however, we can verify that $2Q^\alpha{}_{[\beta\mu]} \equiv \Gamma^\alpha{}_{\beta\mu} - \Gamma^\alpha{}_{\mu\beta}$ transforms as the components of a tensor of $T \otimes T^* \otimes T^*$: $Q^\alpha{}_{[\beta\mu]}$ is the **torsion tensor**.

B.8 Metric connection theories without torsion

The tangent space $T(x)$ can be referred to its natural basis $\{\vec{\partial}_\alpha\}$ or to an arbitrary basis $\{\vec{e}_a\}$.

We introduce in $T(x)$ a scalar product playing a role similar to that introduced into special relativity in T_4 and possessing the same properties. The scalar product is defined by the ten real functions $g_{ab}(x)$ or $g_{\alpha\beta}(x)$, depending on the basis used:

$$\begin{cases} (\vec{V}, \mu\vec{V}_1 + \vec{V}_2) = \mu(\vec{V}, \vec{V}_1) + (\vec{V}, \vec{V}_2) \text{ for } \mu \in K \\ \left(\vec{\partial}_\alpha, \vec{\partial}_\beta\right)_E = \left(\vec{\partial}_\beta, \vec{\partial}_\alpha\right)_E = g_{\alpha\beta}(x) = g_{\beta\alpha}(x) \\ (\vec{e}_a, \vec{e}_b)_E = (\vec{e}_b, \vec{e}_a)_E = g_{ab}(x) = g_{ba}(x). \end{cases} \tag{B.8.1}$$

These relations are satisfied at each point E of the space–time with arbitrary coordinates x.

We verify that g_{ab} and $g_{\alpha\beta}$ transform like the components of a tensor in $T^* \otimes T^*$ giving a fundamental scalar $ds^2 = g_{\alpha\beta}dx^\alpha dx^\beta$ between two events whose coordinates are separated by $\{dx^\alpha\}$. This line element is identified for physical reasons with the quantity $c^2 d\tau^2$, where $d\tau$ is the element of proper time between the two events (cf. II chapter 2).

As an exercise the reader may show that the mathematical framework of special relativity is a special case of the general framework introduced here.

Most generally, K is the complex numbers; $F(x)$ can then describe quantum states of systems. With this aim we introduce into F a scalar product possessing the properties of the quantum-mechanical scalar product.

Einstein's equivalence principle suggests that all the spaces $T(x)$ could be identified with a space with the same structure as T_4 in the neighbourhood of each point of coordinates x_0, as long as expansions are limited to the first order in $x^\mu - x_0{}^\mu$. Under these conditions we postulate the existence of a Minkowski basis of $T(x)$, $\{\vec{h}_a(x)\}$, such that

$$(\vec{h}_a, \vec{h}_b) = \eta_{ab} = \text{diag } (1, -1, -1, -1), \quad a, b, \dots = 0, 1, 2 \text{ or } 3. \quad (B.8.2)$$

The existence of Minkowski bases characterises the scalar product we have introduced.

This kind of basis is called a **tetrad**; in the natural basis it can be decomposed as

$$\vec{h}_a(x) \equiv h^\mu{}_a(x)\vec{\partial}_\mu \text{ with } h^\alpha{}_a h^\beta{}_b g_{\alpha\beta} = \eta_{ab}. \quad (B.8.3)$$

As in special relativity (I § 5.4.4), the existence of a scalar product provides a *fundamental correspondence* between $T(x)$ and its dual $T^*(x)$: in the natural frame indices are lowered or raised using $g_{\alpha\beta}(x)$ or its inverse $g^{\alpha\beta}(x)$. We verify that $g^{\alpha\beta}(x)$ are the components of a tensor of $T(x) \otimes T(x)$ in the natural basis. As in special relativity $g^{\alpha\beta}$, $g_{\alpha\beta}$, $\delta^\alpha{}_\beta$ and $\delta_\alpha{}^\beta$ represent the contravariant, covariant, and mixed components of the fundamental tensor, still called the **metric tensor**. In an arbitrary basis we can use the components $g_{ab}(x)$ or $g^{ab}(x)$ to raise and lower indices.

> As an exercise the reader may show that $h_{ab} \cdot h_\sigma{}^b = g_{\alpha\sigma}$.
>
> In quantum mechanics, K is the complex numbers. The scalar product is not symmetric: $(\vec{V}, \vec{W}) = (\vec{W}, \vec{V})^*$ where the symbol * denotes the complex conjugate. The fundamental correspondence between T and T^* is that between "bras" and "kets". In this case, as $\ell^a{}_{b'}$ may be complex, the transformation of the metric introduced into $F(x)$ is not that of a tensor of $F^*(x) \otimes F^*(x)$ when we change basis.

To ensure the local validity of special relativity it is not sufficient to postulate the existence of Minkowski bases. We must also postulate the existence of a coordinate system such that the connection coefficients vanish at any arbitrarily chosen point x_0. Coordinate systems such that $g_{\alpha\beta}(x_0) = \eta_{\alpha\beta}$ and $\Gamma^\alpha{}_{\beta\mu}(x_0) = 0$ can be locally taken as Minkowski systems of special relativity. To make $\Gamma^\alpha{}_{\beta\mu}(x_0)$ vanish, the torsion introduced above must vanish. As the torsion is a tensor it must vanish in all coordinate systems if $Q^\alpha{}_{[\beta\mu]} = 0$ in any one system. To ensure the correspondence with special relativity, in agreement with the Einstein equivalence principle, we therefore postulate *a priori* the relation

$$\blacklozenge \qquad \Gamma^\alpha{}_{\beta\mu} - \Gamma^\alpha{}_{\mu\beta} = 0. \quad (B.8.4)$$

The scalar product allows us to find observable quantities (cf. I Chap. 4). Thus, when we parallel transport two vectors $\vec{f}_1$ and $\vec{f}_2$ defined at an event A along the same curve, the scalar product must remain fixed. As parallel transport allows one to reproduce identical physical situations elsewhere in space–time the results of measurements must remain unchanged.

The situation is identical in quantum theory. The set of possible results of a measurement is determined by the nature of the measurement. However, the result obtained in a particular measurement is not predetermined, even if the state of the system is known: only the probability of the various possible results is calculable *a priori*. These probabilities are meaningful to us because of the law of large numbers; they are effectively identical to the relative numbers of observed results when the measurement is repeated many times. These probabilities can be expressed by means of a scalar product; it is thus legitimate to assume that the scalar product of two vectors remains unchanged when these are parallel-transported.

In the natural basis this condition can be written

$$\left(\overrightarrow{\mathrm{Im}} \left[\vec{\partial}_\alpha \right], \ \overrightarrow{\mathrm{Im}} \left[\vec{\partial}_\beta \right] \right)_x = g_{\alpha\beta}(x + dx) \tag{B.8.5}$$

where $g_{\alpha\beta}(x) = (\vec{\partial}_\alpha, \vec{\partial}_\beta)_x$ is the scalar product of vectors defined at the point $\{x\}$.

Using (B.8.4), the relation (B.8.5) is equivalent to

$$\blacklozenge \qquad \Gamma^\alpha{}_{\beta\mu} = \frac{1}{2} g^{\alpha\sigma} (g_{\sigma\beta,\mu} + g_{\sigma\mu,\beta} - g_{\beta\mu,\sigma}) \equiv \{^\alpha{}_{\beta\mu}\} = \{^\alpha{}_{\mu\beta}\}. \tag{B.8.6}$$

These relations are also those of special relativity in an arbitrary coordinate system. The right-hand side of the relations (B.8.6) is called the **Christoffel symbol**. In general relativity one thus postulates the equality of the Christoffel symbols and the connection coefficients.

Relation (B.8.5) has the form

$$(\vec{\partial}_\alpha + \Gamma^\sigma{}_{\alpha\mu} dx^\mu \vec{\partial}_\sigma, \ \vec{\partial}_\beta + \Gamma^\rho{}_{\beta\nu} dx^\nu \vec{\partial}_\rho) = g_{\alpha\beta}(x) + g_{\alpha\beta,\mu} dx^\mu$$

where all quantities are evaluated at the point x.

To first order in dx^μ we have $g_{\alpha\beta,\mu} - g_{\tau\beta}\Gamma^\tau{}_{\alpha\mu} - g_{\alpha\tau}\Gamma^\tau{}_{\beta\mu} = 0$.

Cyclically permuting the indices α, β, μ we obtain two equations which we subtract from the previous one:

$$g_{\mu\tau}\Gamma^\tau{}_{\alpha\beta} + g_{\mu\tau}\Gamma^\tau{}_{\beta\alpha} - g_{\mu\alpha,\beta} - g_{\beta\mu,\alpha} + g_{\alpha\beta,\mu} = 0.$$

Contracting each side of this equation with $g^{\mu\rho}$ and using (B.8.4) we obtain the desired relation (B.8.6).

To define the connection in T^*, we assume that the fundamental correspondence between T and T^* defined by the metric tensor is conserved by parallel transport (fig. B.2). The duality of the bases of T and T^* is then conserved; further we get

$$\blacklozenge \qquad \Gamma^{*a}{}_{b\mu} = -\Gamma^a{}_{b\mu} \Rightarrow \Gamma^{*\alpha}{}_{\beta\mu} = -\Gamma^\alpha{}_{\beta\mu}. \tag{B.8.7}$$

A theory conserving the scalar product under parallel transport as here is called a **metric connection theory**.

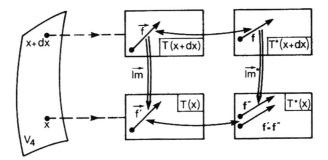

Fig. B.2

Relations (B.8.6) and (B.8.7) show that the covariant derivative of the metric tensor vanishes:

$$\nabla_\mu g_{ab} = 0 = \nabla_\mu g_{\alpha\beta}, \quad \nabla_\mu g^{ab} = 0 = \nabla_\mu g^{\alpha\beta}. \tag{B.8.8}$$

Leibniz's rule allows us to raise and lower indices before or after covariant differentiation; for example

$$\nabla_\mu T_{\alpha\beta} = \nabla_\mu \left(g_{\alpha\sigma} T^\sigma{}_\beta\right) = g_{\alpha\sigma} \nabla_\mu T^\sigma{}_\beta. \tag{B.8.9}$$

Let $\vec{f}_1$ and $\vec{f}_2$ be two vectors of $T(x+dx)$, and let $\mathbf{f}_1$ and $\mathbf{f}_2$ be the corresponding vectors of $T^*(x)$. Let $\vec{f}'_1 = \overrightarrow{\mathrm{Im}}[\vec{f}_1]$ and $\vec{f}'_2 = \overrightarrow{\mathrm{Im}}[\vec{f}_2]$ be the vectors derived from $\vec{f}_1$ and $\vec{f}_2$ by parallel transport from $x + dx$ to x; let $\mathbf{f}'_1$ and $\mathbf{f}'_2$ be the vectors of $T^*(x)$ corresponding to $\vec{f}'_1$ and $\vec{f}'_2$. The definition of the correspondence between T and T^* and the conservation of the scalar product imply

$$\mathbf{f}'_1[\vec{f}'_2] \equiv (\vec{f}'_1, \ \vec{f}'_2) = (\vec{f}_1, \ \vec{f}_2) \equiv \mathbf{f}_1[\vec{f}_2].$$

Let $\mathbf{f}''_1$ be the vector deduced from $\mathbf{f}_1$ by parallel transport of $T^*(x + dx)$ to $T^*(x)$. The condition $\mathbf{f}''_1 = \mathbf{f}'_1$ defines the connection in T^*. In fact let $\{\vec{e}_a\}$ be a basis of $T(x + dx)$ and $\{\mathbf{e}^a\}$ the dual basis of $T^*(x + dx)$, $\mathbf{e}^a[\vec{e}_b] = \delta^a{}_b$. Parallel transporting the two vectors $\vec{e}_a$ and $\mathbf{e}^b$ of $x + dx$ to x in T and T^* respectively, we get

$$\mathbf{e}''^b[\vec{e}'_a] = \mathbf{e}^b[\vec{e}_a] = \delta^b{}_a.$$

The duality of the bases is thus conserved. We use the relations (B.2.2) and (B.6.4) to express $\vec{e}'_a$ and $\mathbf{e}''^b$. To first order in dx^μ we get the relation (B.8.7).

B.9 The notion of curvature

Consider four points of V_4: A_1, A_2, A_3 and A'_2, with respective coordinates x, $x + dx$, $x + dx + \delta x$ and $x + \delta x$. Consider the vector $\vec{f}$ defined in A_3: $\vec{f} = f^a \cdot \vec{e}_a$, $\vec{f} \in F(x + dx + \delta x)$. We displace $\vec{f}$ parallel to itself, following the paths (A_3, A_2, A_1) and (A_3, A'_2, A_1); we obtain the two vectors $\vec{f}_{(\delta d)}$ and $\vec{f}_{(d\delta)}$ both belonging to $F(x)$. In fig. B.3 we have shown the tangent spaces T_1, T_2,

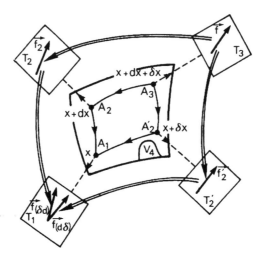

Fig. B.3

T_3 and T'_2 at A_1, A_2, A_3 and A'_2; these are particular spaces F. However, we perform the calculation in the general case.

The relations (B.2.3) allow us to calculate successively $\vec{f}_2$ and $\vec{f}_{(\delta d)}$, then $\vec{f}'_2$ and $\vec{f}_{(d\delta)}$. We find the relations

$$
\blacklozenge \qquad
\begin{cases}
f_{(\delta d)}{}^a - f_{(d\delta)}{}^a = \dfrac{1}{2}(\delta x^\sigma dx^\mu - dx^\sigma \delta x^\mu) \cdot R^a{}_{b\sigma\mu} \cdot f^b \\[2mm]
R^a{}_{b\sigma\mu} = \Gamma^a{}_{b\sigma,\mu} - \Gamma^a{}_{b\mu,\sigma} - \Gamma^a{}_{c\sigma}\Gamma^c{}_{b\mu} + \Gamma^a{}_{c\mu}\Gamma^c{}_{b\sigma}.
\end{cases}
\tag{B.9.1}
$$

$R^a{}_{b\sigma\mu}$ represents the components of the **Riemann–Christoffel tensor**, also called the **curvature tensor**, which belongs to the space $F \otimes F^* \otimes T^* \otimes T^*$. We use the notation $A_{\ldots[ab]\ldots} = 1/2\{A_{\ldots ab\ldots} - A_{\ldots ba\ldots}\}$ to underline the antisymmetry with respect to the indices a and b. We shall sometimes use this notation for the components of the Riemann tensor: $R^a{}_{b[\sigma\mu]} = R^a{}_{b\sigma\mu}$.

> To verify the tensor nature of $R^a{}_{b[\sigma\mu]}$, we could for example use the rules for changes of frames and coordinates for connection coefficients so as to obtain the transformation rule for $R^a{}_{b[\sigma\mu]}$.

When we parallel transport the vector $\vec{f}$ along each of the two paths, the physical state remains "identical to itself"; however, the two final states obtained in general differ from each other if the curvature tensor does not vanish identically. This comes from the fact that the physical meaning given to parallel transport is limited to the first order in dx in the definition of $\overrightarrow{\text{Im}}$. Thus the theories of gravity we consider reduce to special relativity in the neighbourhood of each event x_0 only to the first order in $x^\mu - x_0^\mu$. If we parallel transport a physical system and the measuring apparatus along the same path in the space–time V_4, the results of any measurements on the system with this apparatus will remain the same. The question of knowing what constitutes parallel transport of a physical system in practice remains difficult in general, both in experimental terms and in the design of tests to check the result.

If the curvature tensor vanishes identically, we can show that parallel transport of a vector from a point A to the point B does not depend on the path followed. Two vectors $\vec{f}_1$ and $\vec{f}_2$ defined in x_1 and x_2 derived from each other by parallel transport can be identified with a single vector $\vec{f}$. The spaces $F(x)$ can all be identified with the same space F, which can be one of them for example. We can then choose the bases $\{\mathring{e}_a(x)\}$ so that $\mathring{e}_a(x)$ and $\mathring{e}_a(x')$ correspond through parallel transport. The connection coefficients vanish in this case.

Assume that $\Gamma^a{}_{b\mu}$ is known in a given frame. We seek a frame in which $\Gamma^{d'}{}_{b'\mu}$ vanishes at each point. From (B.6.3) we have

$$\Gamma^{a'}{}_{b'\mu} = \ell^{a'}{}_c \ell^d{}_{b'} \Gamma^c{}_{d\mu} + \ell^{a'}{}_c \partial_\mu \ell^c{}_{b'} = 0.$$

We regard this equation as one determining $\ell^c{}_{d'}$ given $\Gamma^c{}_{d\mu}$.
This equation can be written $\ell^d{}_{b'} \Gamma^c{}_{d\mu} + \partial_\mu \ell^c{}_{b'} = 0$.
The integrability conditions are

$$\partial_{\mu\alpha}(\ell^c{}_{b'}) = \partial_{\alpha\mu}(\ell^c{}_{b'})$$

with $\partial_{\mu\alpha}(\ell^c{}_{b'}) = \partial_\mu(-\ell^d{}_{b'} \Gamma^c{}_{d\alpha}) = \partial_\mu(-\ell^d{}_{b'})\Gamma^c{}_{d\alpha} - \ell^d{}_{b'} \Gamma^c{}_{d\alpha,\mu} = (\ell^a{}_{b'} \Gamma^d{}_{a\mu})\Gamma^c{}_{d\alpha} - \ell^d{}_{b'} \Gamma^c{}_{d\alpha,\mu}$.
We expand $\partial_{\alpha\mu}\ell^c{}_{b'}$ as we expanded $\partial_{\mu\alpha}\ell^c{}_{b'}$. The integrability conditions are then

$$\ell^d{}_{b'} \cdot R^c{}_{d\alpha\mu} = 0 \Leftrightarrow R^c{}_{d\alpha\mu} = 0.$$

When $R^a{}_{b\alpha\mu}$ vanishes, there therefore exist frames at each point such that the connection vanishes. Using these frames equations (B.3.5) have the solution $f^a =$ constant for any path; this demonstrates the claimed property.

Consider the case of special relativity. The connection associated with the tangent space in a Minkowski coordinate system is identically zero, as is the curvature tensor. Conversely, if the curvature tensor associated with the tangent space vanishes, there exists a Minkowski coordinate system. In this case the theory reduces to special relativity in the space–time region where the curvature tensor vanishes, at least if this region is simply connected. The space–time V_4 is then called **flat space**. The presence of gravitation is thus signalled by the fact that the curvature tensor does not vanish identically.

Assume $\Gamma^\alpha{}_{\beta\mu}(x)$ given, and make a coordinate change characterised by $a^\alpha{}_{\mu'} = \partial x^\alpha / \partial x^{\mu'}$. We seek a new coordinate system in which the connection coefficients are zero. The expression (B.7.5) for changes of coordinates for the connection gives the equations $a^\alpha{}_{\beta'} \Gamma^\sigma{}_{\alpha\mu} + \partial_\mu a^\sigma{}_{\beta'} = 0$. As before, the vanishing of the curvature tensor means that solutions $a^\alpha{}_{\beta'}$ of this equation exist. This is not enough to show that $a^\alpha{}_{\mu'}$ is the derivative of the function $x^\alpha(x')$. We must also ensure that the conditions $a^\alpha{}_{\mu',\beta'} = a^\alpha{}_{\beta',\mu'}$ are satisfied. For convenience we consider the inverse matrix $a^{\alpha'}{}_\mu$, of $a^\alpha{}_{\mu'}$. This matrix represents a coordinate change if there

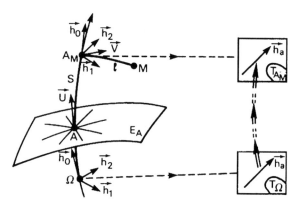

Fig. B.4

exist four functions $x^{\alpha'}(x)$ such that $a^{\alpha'}{}_{\mu} = \partial x^{\alpha'}/\partial x^{\mu}$. The integrability condition here is $a^{\alpha'}{}_{\mu,\beta} = a^{\alpha'}{}_{\beta,\mu'}$.

Using the relation $a^{\beta'}{}_{\alpha} \cdot a^{\sigma}{}_{\beta'} = \delta^{\sigma}{}_{\alpha}$, the starting equation can be written in the equivalent form $\Gamma^{\sigma}{}_{\alpha\mu} = -a^{\beta'}{}_{\alpha}\partial_{\mu}a^{\sigma}{}_{\beta'} = (\partial_{\mu}a^{\beta'}{}_{\alpha})a^{\sigma}{}_{\beta'}$.

The absence of torsion, $\Gamma^{\sigma}{}_{\alpha\mu} = \Gamma^{\sigma}{}_{\mu\alpha}$, implies the required integrability conditions.

Thus in the torsion-free theory we consider, when the curvature tensor vanishes identically there exists a coordinate system such that the connection vanishes also. The relations (B.8.6) imply that the coefficients of the metric tensor are constant in this coordinate system. A coordinate change of the form $x^{\mu'} = A^{\mu'}{}_{\alpha} \cdot x^{\alpha}$, with $A^{\mu'}{}_{\alpha}$ constant, puts $g_{\alpha'\beta'}$ into Minkowski form.

B.10 Locally Minkowski observers

The geodesics introduced in II § 2.3 are the curves satisfying the equations

$$\frac{d^2x^{\mu}}{ds^2} + \{^{\mu}{}_{\alpha\beta}\}\frac{dx^{\alpha}}{ds}\frac{dx^{\beta}}{ds} = 0. \tag{B.10.1}$$

Consider a freely falling observer. This observer falls along a geodesic. Choose a particular event Ω on this geodesic. We fix the points A of the trajectory by the proper length s, measured from Ω. Let $\vec{U}$ be the observer's four-velocity and $\vec{U}/c$ the unit vector of $T(x)$ tangent to the trajectory at the point A. Consider the set of geodesics from A and orthogonal to $\vec{U}$ at the point A. These curves form a three-dimensional space, called E_A (fig. B.4).

At Ω we choose a tetrad $\{\vec{h}_a\}$ whose vector $\vec{h}_0$ is the same as $\vec{U}/c$ at this point, $\dfrac{\vec{U}}{c} = \vec{h}_0$ and $(\vec{h}_a, \vec{h}_b) = \eta_{ab}$. We parallel transport the tetrad vectors along the observer's worldline. We thus construct a frame in $T(x)$ at each point A. This frame is a tetrad whose vector $\vec{h}_0$ is tangent to the observer's trajectory.

As an exercise the reader can show that a unit tangent vector to a geodesic remains tangent to the geodesic under parallel transport along it. The conservation of the scalar product then demonstrates the property proved here.

Now consider an arbitrary point M in the neighbourhood of the trajectory. This point belongs to a particular three-space E_A. Let A_M be the corresponding point A and x^0 the associated value of s at A_M. The geodesic of E_{A_M} passing through M has at A_M the unit tangent vector $\vec{V}$, such that $(\vec{V}, \vec{V}) = -1$. We decompose $\vec{V}$ in the tetrad at A_M: $\vec{V} = V^a \cdot \vec{h}_a$. Let ℓ be the proper length *along the geodesic* from A_M to M. We then set

$$x^1 = \ell \cdot V^1, \quad x^2 = \ell \cdot V^2, \quad x^3 = \ell \cdot V^3, \quad x^0 = s \qquad \text{(B.10.2)}$$

with $\ell = \int_{A_M}^M d\ell$ and $d\ell = \sqrt{-g_{\alpha\beta} dx^\alpha dx^\beta}$.

We have thus constructed a coordinate system such that $g_{\alpha\beta} = \eta_{\alpha\beta}$, $g_{\alpha\beta,\mu} = 0$ and $\Gamma^\alpha{}_{\beta\mu} = 0$ at each point A of the goedesic considered. This is called a **Fermi coordinate system** associated with a freely falling observer. This system is locally Minkowskian in the neighbourhood of the observer's worldline. According to the Einstein equivalence principle, in this system physical laws take the same form as in a Minkowski system in special relativity. This system can be identified with a Galilean system in Newtonian theory for the description of experiments in the observer's neighbourhood, as long as the approximation of small velocities holds. The tetrads we have introduced constitute a **local inertial frame**.

As an exercise one can show that this coordinate system has the claimed properties: $g_{\alpha\beta} = \eta_{\alpha\beta}$, $g_{\alpha\beta,\mu} = 0$, $\Gamma^\alpha{}_{\beta\mu} = 0$. Show that the vectors $\vec{h}_a$ with components $\delta^\mu{}_a$ in the natural basis associated with Fermi coordinates are given by parallel transport along the curve $x^k = 0$, and that the curve with equation (B.10.2) with s = constant is a geodesic.

B.11 Special tensors and identities

The space–time we have introduced is a space with a metric connection, without torsion, and locally Minkowskian; this is a **pseudo-Riemannian space**. It can be studied as a geometrical object, but in this section we shall limit ourselves to the study of several tensors of particular physical interest. We shall give a few properties whose proofs either are immediate deductions from the definitions, or can be found in books on differential geometry.

We have emphasised the importance of the curvature tensor $R^\alpha{}_{\beta[\mu\sigma]}$, whose vanishing characterises the absence of gravity. This tensor has the following symmetries:

$$\begin{cases} R^\alpha{}_{\beta\mu\sigma} \equiv -R^\alpha{}_{\beta\sigma\mu} \equiv R^\alpha{}_{\beta[\mu\sigma]} \\ g_{\alpha\tau}R^\tau{}_{\beta[\mu\sigma]} \equiv R_{\alpha\beta[\mu\sigma]} \equiv R_{[\alpha\beta][\mu\sigma]} \equiv R_{[\mu\sigma][\alpha\beta]} \\ R^\alpha{}_{\beta[\mu\sigma]} + R^\alpha{}_{\sigma[\beta\mu]} + R^\alpha{}_{\mu[\sigma\beta]} \equiv 0. \end{cases} \qquad \text{(B.11.1)}$$

In the usual notation, $( )_{;\mu} = \nabla_\mu( )$, we have

$$R^\alpha{}_{\beta[\mu\sigma];\tau} + R^\alpha{}_{\beta[\tau\mu];\sigma} + R^\alpha{}_{\beta[\sigma\tau];\mu} \equiv 0. \tag{B.11.2}$$

These are the **Bianchi identities**.

The covariant derivative operator ∇_μ is a generalisation of the partial derivative operator ∂_μ. These operate in the same way on scalar fields. They act in the same way on components of a tensor field at any point where $g_{\alpha\beta,\sigma} = 0$.

However, unlike partial derivatives, covariant derivatives do not in general commute:

$$(\nabla_\mu \nabla_\tau - \nabla_\tau \nabla_\mu)[A^{\alpha \cdots}{}_{\beta \dots}] \equiv - R^\alpha{}_{\sigma[\mu\tau]} A^{\sigma \cdots}{}_{\beta \dots} - \dots$$
$$+ R^\sigma{}_{\beta[\mu\tau]} A^{\alpha \cdots}{}_{\sigma \dots} + \dots \tag{B.11.3}$$

The **Ricci tensor**, $R_{\beta\mu}$, is obtained by contraction of the curvature tensor, while contraction of the Ricci tensor gives the **scalar curvature** R:

$$R_{\beta\mu} \equiv R^\alpha{}_{\beta[\mu\alpha]} \equiv R_{\mu\beta} \qquad R \equiv R^\alpha{}_\alpha \equiv g^{\alpha\beta} R_{\beta\alpha}. \tag{B.11.4}$$

The **Einstein tensor** is defined from the previous quantities:

$$S_{\alpha\beta} \equiv S_{\beta\alpha} \equiv R_{\alpha\beta} - \frac{1}{2} g_{\alpha\beta} R. \tag{B.11.5}$$

From the Bianchi identities we deduce the identity

$$S^{\alpha\beta}{}_{;\beta} \equiv 0. \tag{B.11.6}$$

Using the symmetries of the curvature tensor and contracting α and σ in (B.11.2), we get $R_{\beta\mu;\tau} + R^\alpha{}_{\beta\tau\mu;\alpha} - R_{\beta\tau;\mu} \equiv 0$.
Contracting this identity with $g^{\beta\mu}$ gives $R_{;\tau} - R^\alpha{}_{\tau;\alpha} - R^\alpha{}_{\tau;\alpha} \equiv 0$.
The latter identity gives the identity (B.11.6).

We define similarly the **Weyl tensor**

$$C_{[\alpha\beta][\mu\tau]} \equiv R_{\alpha\beta\mu\tau} + \frac{1}{2}(g_{\alpha\mu}R_{\beta\tau} - g_{\alpha\tau}R_{\beta\mu} - g_{\beta\mu}R_{\alpha\tau} + g_{\beta\tau}R_{\alpha\mu}) - \frac{1}{6}R(g_{\alpha\mu}g_{\beta\tau} - g_{\alpha\tau}g_{\beta\mu})$$

This tensor has the same symmetries as the curvature tensor; it also satisfies the relation $C^\alpha{}_{\beta\mu\alpha} \equiv 0$. The Weyl tensor vanishes iff the metric is **conformal** to a flat-space metric; i.e. iff there exists a coordinate system such that $g_{\alpha\beta} = f(x) \cdot \eta_{\alpha\beta}$ where $f(x)$ is a scalar function.

As an exercise, verify that the Weyl tensor vanishes for the cosmological solutions studied in chapters 5 and 6 of part II.

Useful numerical values

Mathematical constants

$\pi = 3.1415926535\ldots$

$e = 2.7183 = 10^{0.43429}$, $\log e = 0.43429$

$10 = e^{2.3026}$, $\ln 10 = \log_e 10 = 2.3026$

$1° = 1.7453 \times 10^{-2}$ rd, $\quad 1' = 2.9089 \times 10^{-4}$ rd, $\quad 1'' = 4.8481 \times 10^{-6}$ rd

Physical constants

gravitational constant:
$$G = 6.673 \times 10^{-8} \text{ cgs}, \qquad G = 6.673 \times 10^{-11} \text{ m}^3 \text{ kg}^{-1} \text{ s}^{-2}$$

Coulomb constant : $\qquad\qquad\qquad 1/4\pi\varepsilon_0 = 8.987 \times 10^9 \text{ m}^3 \text{ kg C}^{-2} \text{ s}^{-2}$

magnetic constant: $\qquad\qquad\qquad \mu_0/4\pi = 1.000 \times 10^{-7} \text{ m kg C}^{-2}$

speed of light in vacuo:
$$c = 2.998 \times 10^{10} \text{ cgs}, \qquad c = 2.998 \times 10^8 \text{ m s}^{-1},$$

Planck's constant:
$$\hbar = 1.055 \times 10^{-27} \text{ cgs}, \qquad \hbar = 1.055 \times 10^{-34} \text{ J s}$$
$$\hbar = 6.582 \times 10^{-16} \text{ eV s}, \qquad h = 2\pi\hbar = 6.625 \times 10^{-34} \text{ J s}$$

Boltzmann's constant :
$$k_B = 1.381 \times 10^{-16} \text{ cgs}, \qquad k_B = 1.381 \times 10^{-23} \text{ J K}^{-1}$$

elementary charge : $\qquad\qquad\qquad e = 1.602 \times 10^{-19} \text{ C}$

electron mass :
$$mc^2 = 0.511 \text{ MeV}, \qquad m = 9.1096 \times 10^{-31} \text{ kg}$$

proton mass:
$$M_P c^2 = 938.3 \text{ MeV}, \qquad M_P = 1.673 \times 10^{-27} \text{ kg}$$

neutron mass:
$$M_n c^2 = 939.6 \text{ MeV}, \qquad M_n = 1.675 \times 10^{-27} \text{ kg}$$

Bohr radius:
$$a = 0.5292 \text{ Å}, \qquad a = 5.292 \times 10^{-11} \text{ m}$$

Astronomical quantities

year: $\qquad$ 1 y$=3.16 \times 10^7$ s

light year:

$\quad$ 1 ly$= 9.46 \times 10^{17}$ cm, $\qquad$ 1 ly$=9.46 \times 10^{15}$ m

Earth–Sun distance (astronomical unit):

$\quad$ 1 AU$= 1.50 \times 10^{13}$ cm, $\qquad$ 1 AU$=1.50 \times 10^{11}$ m

parsec:

$\quad$ 1 pc$= 3.09 \times 10^{18}$ cm, $\qquad$ 1 pc$=3.09 \times 10^{16}$ m

$\quad$ 1 pc$= 3.26$ ly

Hubble constant: $\qquad H=75$ km s^{-1}/Mpc$=2.43 \times 10^{-18}$ s^{-1}

Hubble time: $\qquad T_H=1/H = 13 \times 10^9$ y$=4.12 \times 10^{17}$ s

apparent luminosity, l, as a function of apparent bolometric magnitude, m:

$\quad l= A \times 10^{-0.4\ m}$

$\quad A = 2.52 \times 10^{-5}$ erg s^{-1} cm^{-2}, $\qquad A=2.52 \times 10^{-8}$ W m^{-2}

absolute luminosity L, as a function of absolute bolometric magnitude, M:

$\quad L= B \times 10^{-0.4\ M}$

$\quad B = 3.02 \times 10^{35}$ erg s^{-1}, $\qquad B=3.02 \times 10^{28}$ W

Sun

mass: $M = 1.99 \times 10^{33}$ g, $\qquad M=1.99 \times 10^{30}$ kg

radius: $\qquad R=6.96 \times 10^5$ km

Schwarzschild radius : $\qquad 2GM/c^2=R_s = 2.95 \times 10^3$ m

absolute luminosity:

$\quad L= 3.90 \times 10^{33}$ erg s^{-1}, $\qquad L=3.90 \times 10^{26}$ W

angular momentum: $J/Mc = 280$ m, $\qquad J=1.7 \times 10^{41}$ kg m^2 s^{-1}

Earth

mass: $M= 5.98 \times 10^{27}$ g, $\qquad M=5.98 \times 10^{24}$ kg

equatorial radius: $\qquad R=6.38 \times 10^3$ km

Schwarzschild radius: $\qquad 2GM/c^2=R_s = 8.86 \times 10^{-3}$ m

mean orbital velocity: $\qquad v=29.8 \times 10^3$ m s^{-1}

Moon

mass: $M= 7.35 \times 10^{25}$ g, $\qquad M=7.35 \times 10^{22}$ kg

lunar radius: $\qquad R=1.738 \times 10^3$ km

Schwarzschild radius: $\qquad 2GM/c^2=R_s = 1.09 \times 10^{-4}$ m

mean Earth–Moon distance: $\qquad d=3.84 \times 10^8$ m

Planets of the Solar System

Planet	Period (years)	L (10^6 km)	Eccentricity
Icarus	1.12	51.0	0.827
Mercury	0.241	55.5	0.206
Venus	0.615	108	0.00682
Earth	1.00	150	0.0167
Mars	1.88	226	0.0933
Jupiter	11.9	776	0.0483
Saturn	29.5	1420	0.0559
Uranus	84.0	2860	0.0471
Neptune	165	4500	0.0085
Pluto	248	5500	0.2494

$2/L = 1/r_+ + 1/r_-$, where $r_\pm$ are the extremal distances of the planet from the Sun

Some clusters of galaxies

Cluster	Estimate of number of galaxies in the cluster	cZ (km s^{-1})
Virgo	2500	1150
Cancer	150	4800
Perseus	500	5400
Coma	1000	6700
Ursa Major I	300	15400
Leo	300	19500
Ursa Major II	200	41000

Constellations correspond to visual appearances, and have no physical significance. They specify a set of directions in the sky.

Bibliography

This bibliography is deliberately limited as there are many books on this area. If books complementing the present one are required we recommend those marked with asterisks.

Special relativity

M. Born, *Einstein's Theory of Relativity*, Dover Publications Inc., New York, 1962.
> Detailed presentation of special relativity in a neo-Newtonian spirit; some general relativity (about 370 pages).

J. L. Synge, *Relativity: the Special theory*, North-Holland Publishing Company, Amsterdam, 1956.
> Difficult, but many important theoretical questions are discussed (about 450 pages).

General relativity

L. Landau and E. Lifshitz, *Classical Theory of Fields*, Pergamon Press, Oxford, 1985.
> Purely theoretical presentation of special and general relativity and classical electromagnetic theory; some interesting, often difficult exercises (about 450 pages).

C. W. Misner, K. Thorne, J. A. Wheeler, *Gravitation*, W. H. Freeman and Company, San Francisco, 1973. (*)
> This book is very complete, and very interesting in many ways. For aspiring physicists as well as experts in the field (about 1300 pages).

B. F. Schutz, *A First Course in General Relativity*, Cambridge University Press, 1985.
> This book is mainly devoted to general relativity, and contains tractable exercises and an annotated bibliography (about 400 pages).

R. d'Inverno, *Introducing Einstein's Relativity*, Clarendon Press, Oxford, 1993.

Experimental gravitation

C. Will, *Theory and Experiment in Gravitational Physics*, Cambridge University Press, Cambridge, 2nd edn, 1993.
> This slightly austere book discusses the main experimental and observational results in the framework of parametrised post-Newtonian theories (about 350 pages).

Rendiconti della Scuola Internazionale di Fisica "Enrico Fermi". Academic Press, London.
XX Corso : *Verification of Gravity Theories*, C. Møller (Ed.), 1962.
LVI Corso : *Experimental Gravitation*, B. Bertotti (Ed.), 1974. (**)
> In the first of these books the article by R. Dicke on Mach's principle and the equivalence principle is well worth close attention. In the second book the article by C. Will presents the parametrised post-Newtonian formalism in an accessible way. There are also some articles on gyroscopic satellites.

Cosmology

S. Weinberg, *Gravitation and Cosmology*, John Wiley & Sons Inc., New York, 1972. (***)
> Interesting physical and often concrete approach to gravitation, cosmology and astrophysics (about 650 pages).

S. Weinberg, *The First Three Minutes*, Basic Books, New York, 1977.
> Well-written popular book.

J. V. Narlikar, *Introduction to Cosmology*, Cambridge University Press, 1993.
> This book gives a modern treatment of theoretical and observational cosmology.

History of relativity

C. C. Gillespie, *Dictionary of Scientific Biography*, Charles Scribner's Sons, New York, 1970.
> This dictionary gives biographical details of physicists and mathematicians quoted in the present book.

The Principle of Relativity, Dover Publications Inc., New York, 1923.
> Collection of papers constituting important steps towards general relativity from 1895 to 1918.

C. W. Kilmister, *Special Theory of Relativity*, Pergamon Press, Oxford, 1970.
> Succinct presentation of special relativity by the author, and a collection of important papers on the subject between 1881 and 1937.

J. Eisenstaedt, "The early interpretation of the Schwarzschild solution" and "The low water-mark of general relativity 1925–1955" in *Einstein and the History of General Relativity*, proceedings of the 1986 Osgood Hill

Conference. Einstein studies, vol. I, J. Stachel and D. Howard (Eds.), Birkhäuser, Boston, Mass., 1989 (pp. 213–233 and pp. 277–292).

> These articles are very accessible; they concern the history of science and epistemology.

Supplementary

Three Hundred Years of Gravitation, S. W. Hawking and W. Israel (Eds.), Cambridge University Press, 1987.

> A collection of frequently difficult articles, which nevertheless give a good impression of the questions studied in the 1980s. Contains an article by K. Thorne on interferometric detection of gravitational waves.

B. Schutz, *Geometrical Methods of Mathematical Physics*, Cambridge University Press, 1980.

Index